Technology of the Guitar

T0180942

Richard Mark French

Technology of the Guitar

Foreword by Bob Taylor, President, Taylor Guitars

 Springer

Richard Mark French
Department of Mechanical Engineering Technology
Purdue University
West Lafayette, IN 47907
USA

FENDER®, STRATOCASTER®, TELECASTER®, STRATACOUSTIC™, T-BUCKET™, ADAMAS®,
and the distinctive headstock designs commonly found on these guitars are the trademarks of Fender Musical
Instruments Corporation and used herein with express written permission. All rights reserved.

ISBN 978-1-4899-8790-7   ISBN 978-1-4614-1921-1 (eBook)
DOI 10.1007/978-1-4614-1921-1
Springer New York Heidelberg Dordrecht London

© Springer Science+Business Media New York 2012
Softcover reprint of the hardcover 1st edition 2012

This work is subject to copyright. All rights are reserved by the Publisher, whether the whole or part of
the material is concerned, specifically the rights of translation, reprinting, reuse of illustrations,
recitation, broadcasting, reproduction on microfilms or in any other physical way, and transmission or
information storage and retrieval, electronic adaptation, computer software, or by similar or dissimilar
methodology now known or hereafter developed. Exempted from this legal reservation are brief excerpts
in connection with reviews or scholarly analysis or material supplied specifically for the purpose of being
entered and executed on a computer system, for exclusive use by the purchaser of the work. Duplication
of this publication or parts thereof is permitted only under the provisions of the Copyright Law of the
Publisher's location, in its current version, and permission for use must always be obtained from
Springer. Permissions for use may be obtained through RightsLink at the Copyright Clearance Center.
Violations are liable to prosecution under the respective Copyright Law.
The use of general descriptive names, registered names, trademarks, service marks, etc. in this
publication does not imply, even in the absence of a specific statement, that such names are exempt
from the relevant protective laws and regulations and therefore free for general use.
While the advice and information in this book are believed to be true and accurate at the date of
publication, neither the authors nor the editors nor the publisher can accept any legal responsibility for
any errors or omissions that may be made. The publisher makes no warranty, express or implied, with
respect to the material contained herein.

Cover illustration: Cover image courtesy of Lindsay Madonna Konrady

Printed on acid-free paper

Springer is part of Springer Science+Business Media (www.springer.com)

Foreword

My friend Sharon makes a good cookie. She built a kitchen to make good cookies. She does it every day and she even passes out tubs of her frozen cookie dough to friends. People eat her cookies and say, "Hmmm, that's a good cookie, " and go on with their day. They don't realize how much time she's put into it or that her home kitchen could convert into a business if she chose.

Sharon knows that butter needs to be in her cookie dough and it has to be cold butter. Your grandma probably knows that too. Let the butter warm up and you can just forget that batch, it won't be good. I never knew that. I wanted some flourless, low sugar cookies one day, so I got out my oatmeal and stabbed away. You could eat what I made, but they fell apart, and in the end, they didn't qualify as a cookie. I tried commercial versions of my dream cookie and they all contained ingredients that I didn't want to eat. I asked Sharon to give it a whirl.

As I said, Sharon knows how to make a cookie, and so she returned to me a plate of wonderfully made, delicious, healthy cookies that used all the ingredients I liked. She knew that there had to be some basic ingredients to make it work, so there was some sugar and some butter, and some other total essentials. That's fine. Without them it wouldn't be a cookie, it would be something else, but she achieved the main goal while I failed at it. Given the challenge she employed what she knows to be true, and I was totally impressed!

She's not a chemist.

A chemist could come back and give us all the reasons why cold butter does the trick, and why some sugar, rather than another sweetener, like Stevia, will help hold it all together. They might even define what a cookie is in terms that the normal cookie-eater wouldn't understand. But the chemist is still right; as right as Sharon, and if they spent an evening together combining their knowledge they'd find common ground and interest over a cookie. They'd probably end up being friends.

I know how to make a guitar, like Sharon knows how to make a cookie. I'm not scientific about it. I'm logical, and one might consider that scientific. I've heard some people explain the Scientific Method as having a theory, testing the theory, and then believing the results of the test. Believing the results is the hard part because we want our theories to be correct, as we're very emotional about our theories. They're like our little babies, and we want them to live.

But, alas, sometimes the testing shows that the theory isn't right at all, and if one can believe those results and move on, they end up learning, and amassing knowledge. I must say that I'm pretty good at that, and so in that small way I would say I'm scientific about my approach to guitar building.

But I'm not scientific in the same way that Mark French is. Just open these pages and you'll see what I mean. Now, if you're a real scientist, or a mathematician, or an engineer, you'll open the pages and feel right at home. If you're like me, you'll open them and get scared! I mean, just one of the equations makes me feel stupid. But if you're like me, happily working away in your shop with the northern light filtering in through your favorite window, and you like the smell of coffee when you're shaping your braces, don't close this book too fast. Let me tell you why.

Mark French has delved deep into things about the guitar, that people like me couldn't conceive to do in the first place. And by reading this, one can corroborate things they've thought to be true, or possibly prevent themselves from moving down a track of thought that has no merit. Knowledge is knowledge and when someone like Mark puts this much work on paper, presented so clearly, we should all stand up and clap.

Let me give you an example. Every acoustic guitar maker is faced with the task of putting a pickup on the guitar they make and trying to find the best location. We all think there is a "sweet spot" on the guitar and if we could find it, then it would be the solution to our problem. But it turns out that the sweet spot is kind of like the pot of gold at the end of the rainbow. What is a sweet spot for one frequency is very likely to be a node point for another frequency. Mark proves it with motion photography. When a guy like me sees this real evidence, it's incredibly helpful because we can quit prospecting in the wrong tunnel for our mother lode, because it's not there!

When I buy a new car, I don't open the owner's manual the first day. I just drive. But days come during the life of that car when I do open it and learn something new. Sometimes it's shocking news, like, I had no idea my car could have the volume of the radio set to increase with speed automatically, to counteract the increased noise! Wow!

That info was there on day one, but I wasn't in what I like to call a teachable moment. When I look at Mark's book there are so many topics, so deeply explored that I just know that a day here and a day there will come where I'll pull this book from the shelf when I'm in the middle of trying to understand something to a greater depth.

Meanwhile, Mark calls me, or others at Taylor Guitars and asks questions about guitars as though we are experts. Why? Because we are, we're experts at making guitars like Sharon is at making cookies. We actually do it, and sell them, and listen to people play them. We know what happens to the sound when we make changes. So together, folks like me and those like Mark build a broader understanding of what is actually happening. I love that, and am happy that Mark is so interested in this topic of guitars that he pours this much detailed time into it. I won't do it, it's not my nature, so I'm very grateful to Mark French for doing the massive amount of work needed to compile this book. It belongs on the shelf near the workbench, for easy access.

El Cajon, CA, USA   Bob Taylor

Preface

This is the second book I've done on the technical aspects of guitar. Why a second? There are a couple reasons. One is that the first book, Engineering the Guitar, uses some advanced math that discouraged many otherwise interested readers. Some of these people are teachers at high schools and community colleges who are using guitar making as a means of interesting their students in technical subjects. There was a lot of interest in a book that used math no more advanced than junior high school algebra.

Another reason is simply that I've learned a lot since doing the first book. I'm incredibly fortunate that luthiers all over the world have been willing to share their knowledge with me. As a result, I've gotten smarter about what goes into making guitars.

It is my hope that professional luthiers, teachers, hobbyists and people still thinking about making their first instrument will find useful information in this book. I've tried very hard to make it easy to read while still being informative and, above all, useful. You'll have to decide whether I succeeded.

West Lafayette, IN, USA   Richard Mark French

Making Guitars – usually the best thing to be doing

Acknowledgements

It's hard to know where to start in acknowledging all the help I've gotten in writing this book. I've found the community of guitar makers to be interesting, creative, generous, inspiring and more than a little quirky – definitely the kind of folks with whom one should associate. As I read back over the first draft of this section, I couldn't help noticing that I have described people in glowing terms – as if they are all the best things since sliced bread. Well, they are and it has been my great fortune to meet them.

I've simply lost track of how many people have made helpful suggestions, contributed pictures and shared their knowledge with me. I've tried very hard to acknowledge sources of images and reference the work of others whenever I've used it. My profound apologies to those who I have overlooked.

Even among the group of people who have helped me as I wrote this book, there are some who stand out. Tim Shaw has been more than gracious in sharing from his vast stock of knowledge about all things guitar and has gone far beyond any reasonable expectation in supporting my students. He has also been encouraging and unfailingly upbeat. Josh Hurst has shared more tips on how to make guitars than I can count. As a graduate of this department (before I arrived – alas, I can claim no credit for his success), he has been an inspiration for many students who have followed.

Bob Taylor, Dave Hosler and the gang at Taylor Guitars have been great and have definitely lived up to Bob's vision of leaving the guitar in a better state than he found it. Their openness and willingness to share information sets a standard that few companies can match. That the company is thriving shows that nice folks don't always finish last.

Dick Boak has been a great help, providing images for the book, detailed information on Martin instruments and historical information I would have had a hard time finding any other way. He, along with his co-authors, literally wrote the book on Martin Guitars. Most people don't know that he's also a skilled artist. It's good to know that Renaissance personhood still lives.

Tim Olsen at the Guild of American Luthiers and R.M. Mottola, GAL person and an expert on technical aspects of the guitar, have been e-mail friends until this summer, when I finally met them in person. They have both been great sources of information and have introduced me to many creative and accomplished luthiers. They also seem to think that making guitars is just the best thing ever. What can I do but agree?

Richard Bruné, a very skilled builder and player, is proof that an autodidact can be a scholar of a high order. He has graciously shared his knowledge, both of making fine guitars and of their history, and has taught me to aim high. He also let me play some seriously nice guitars. What was he thinking?

Mike Jacob has patiently continued the education of this aerospace guy as I try to learn electronics. It is no mystery that he is known around here as a great teacher. I wonder if he'll eventually get tired of answering my questions. I hope not.

Kevin Beller knows as much about designing guitar pickups as anyone and has unhesitatingly shared his knowledge, both with me and with my students. He's also supported my students beyond anything I dared hope. I and hundreds of new luthiers are in his debt.

Brad Harriger has been my partner in running guitar workshops here for a while now and together, we have learned a lot about how to make guitars by the dozen. He also has a classic case of guitar acquisition syndrome. I fear he might have caught it from me.

I owe a special thanks to The Minions: Craig Zehrung, Jim Stratton and Patrick Morrison. You are bright, creative, industrious and endlessly entertaining. It is my happy challenge to live up to your expectations; may all professors have such students. I suppose I'll have to let you graduate eventually.

Dave Brantley, one of the best elementary school teachers ever, has listened to crazy ideas, offered some of his own and showed me how to strike a spark of learning in some very small luthiers. I've gotten to watch a master introduce the joy of learning to his young students. It's been a real treat to see the first glow of fires that need never go out. I wish I could buy stock in those kids.

Kay Solomon has been an unfailing source of encouragement over the course of writing this book. She has inspired and cared for the students here for a long time. Most of them know how fortunate they are to have her on their side and few of them hesitate to tell me about her general fabulosity. I couldn't agree more. For some inexplicable reason, she seems to like editing and has graciously line edited this manuscript. Any remaining errors are certainly mine only. Kay, now it's your turn to write a book.

I must offer special thanks and a mea culpa to my wonderful, long-suffering wife, Amy. You told me at the beginning of this project that it was going to be too much work and then spent many months watching me prove you right. I promise, no more books for a while.

Finally, I want to thank Brian and Kate. They seem to have accepted that their Dad has the bug to write books and have been willing to humor me. Brian, I hope you find there is always just one more guitar to make. Kate, I hope you will continue to serenade me with your baritone horn whenever you sense a disturbing absence of low brass in my life.

Contents

Chapter 1
Overview

The best makers are the ones in touch with the market, with an open mind and with experience and historical knowledge to draw from. While copying the great guitars of the past is an evolutionary dead end, trying to make the great guitars of the future in the absence of knowing what has preceded them is also futile.

R.E. Bruné

More than many other modern implements, musical instruments are strongly influenced by tradition. Most modern classical guitars strongly resemble those from 100 years ago and some steel string acoustic guitar designs have been in production for more than 70 years. While luthiers (makers of stringed instruments, like guitars) are a constant source of innovation, a strong current of tradition underlies most of the industry, from individual builders to large manufacturers.

Understanding the history of guitars can help explain some of their design features. It's also good for the prospective builder to understand where he or she would like to fit in. Additionally, if one wants to make historically accurate reproductions, as some makers do, then it is necessary to have a basic idea of what kinds of instruments are historically legitimate.

Before moving on, however, we need to have a language to describe the parts of the guitar. Figure 1.1 shows a guitar with the parts labeled. This instrument is a classical guitar – an acoustic guitar with nylon strings. However, the terminology is basically the same for other kinds of guitars.

1.1 Early Stringed Instruments

The history of the guitar is a long and rich one. Stringed musical instruments seem to have been part of human culture for at least as long as there have been written records. Certainly, there is clear evidence of stringed instruments being used more than 4700 years ago [1]. At least two lyres were found in the archeological excavations at Ur, an ancient Sumerian city in what is now Iraq.

R.M. French, *Technology of the Guitar*, DOI 10.1007/978-1-4614-1921-1_1,
© Springer Science+Business Media New York 2012

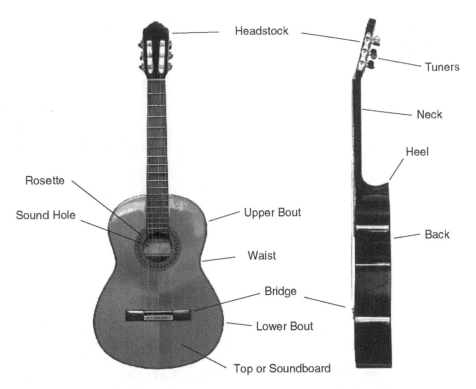

Fig. 1.1 Classical Guitar with Parts Labeled (Guitar image from Wikimedia Commons, image is in the public domain)

The lyre is an instrument with strings and a small hollow body, but it is related to the guitar in only the most remote way. To be a true antecedent of the guitar, an instrument should at least have a clearly defined neck and body. An Egyptian painting, dating from 1422-1411 BCE, shows musicians playing a harp and a long-necked instrument resembling the more modern tanbur (Fig. 1.2). This necked instrument is more suggestive of the European family of necked, stringed instruments that developed into the guitar.

It would be nice if the history of the guitar was a tidy succession of well-defined instruments. Alas, this is not at all the case. It was preceded by whole families of stringed instruments developed by a number of different cultures. To add yet another complication, the guitar coexisted with other instruments that bore a resemblance to it. The history of stringed instruments, though convoluted, hints at the extent to which they have long been a part of human societies around the world.

It's clear that many early civilizations around the world developed stringed instruments [2]. Certainly, they were used in China, India, Africa and the Middle East. From such a diverse background, there is no single clear historical timeline that cleanly leads us to the modern guitar. There have been overland trade routes for millennia and trading across the globe by sea has been possible for at least

Fig. 1.2 Painting of Egyptian Musicians, 1422-1411 BCE and a Modern Tanbur (Wikimedia Commons, images are in the public domain)

500 years. There is, thus, every reason to assume that geographically separated cultures were aware of one another's musical instruments just as they were of spices, fabrics and other goods. Stringed instruments are usually compact and easy to transport; it is easy to imagine that they were tucked into the packs and holds of travelers wanting some music to ease the monotony of long journeys.

Mapping the complete lineage of the guitar is well beyond the scope of this book and some good histories of the instrument already exist [3, 4, 5]. It's enough here to acknowledge that the early history of stringed instruments includes many different types from many disparate societies around the world. That said, the guitar is essentially a European instrument.

Simply tracing the origins of the word guitar is helpful in showing how many different influences there may have been in the instrument's development. The Old Persian word *sihtar* became the Greek work *kithara* [6]. *Tar* means string in Persian and there are a number of stringed instruments with the word *tar* in the name. Perhaps the most familiar of these now, other than the guitar, is the sitar from south Asia.

A kithara (Fig. 1.3) is similar to a lyre and is small enough to be easily portable. However, the fact that the word *kithara* sounds a bit like the word *guitar* is hardly enough evidence to establish the kithara as a precursor to the guitar.

The Greek work *kithara* led to the Latin *cithara*. In the Andalusian dialect of Arabic, the word is *qitara*. Andalusia (Al-Andalus) was the Arabic name for the parts of the Iberian peninsula and what is now southwest France that was then ruled by the Moors (Arab and North African Muslims). The Moors were eventually

Fig. 1.3 Apollo Holding
a Kithara, 2nd Century CE
(Wikimedia Commons,
image is in the public domain)

expelled from what is now Spain and the Spanish word *guitarra* appeared. This became *guitar* in English, *guitare* in French, *chitarra* in Italian and *Gitarre* in German. Thus, the derivation of the name suggests an instrument that started in Persia and Greece – essentially the same places that western civilization started – and followed the development of culture and learning to Europe.

It's somewhat confusing that the name guitarra and its derivatives evolved separately from the instrument that is now called the guitar. The name was given to instruments that were not what we now consider to be direct precursors to the modern guitar. The Kithara shown in Fig. 1.3 has only the most distant relationship to the modern guitar; it has strings and a small, enclosed body chamber, but no neck and no frets [7]. This situation existed until the beginnings of the Renaissance. Indeed, in the 15th century, the name referred to high pitched members of the lute family. Not until the 16th century did the guitar exist as a distinct instrument [8].

Perhaps the most familiar stringed instrument from the late middle ages and early Renaissance is the lute. There are some superficial similarities between it and the guitar, but it is not necessarily the direct precursor to the guitar that it might seem to be.

The lute has a pear-shaped body with a deep, rounded back. There is a short, wide neck and a headstock mounted at a sharp angle to the neck. The name lute is derived from the name of the Arabic instrument, the oud, whose name, in turn, came from the Arabic *al 'ud* (literally "the wood"). The oud is very similar to the

Fig. 1.4 A Modern Replica of Renaissance Lute (Image reproduced courtesy of Daniel Larson, http://www.daniellarson.com)

European lute with the most obvious difference being a narrower neck. The very name luthier, which now refers to any maker of stringed instruments, is derived from the word lute [9].

Lutes are typically strung with pairs of strings called courses. Lutes usually have one sound hole filled with an elaborate decoration and placed on the centerline of the instrument. Figure 1.4 shows a nice Renaissance lute made by luthier Daniel Larson.

The lute certainly influenced the development of the guitar, but it kept its identity as a separate instrument. Its design evolved, however, with the demands of new styles of music. Medieval lutes had four or five courses of strings and were generally played with a quill. During the Renaissance, the number of strings was increased to six courses or more and it became common for the player to pluck the strings with fingertips. The instrument remained popular up to the Baroque era. While now a more obscure instrument, it still has a following and there are active players and societies promoting it.

One of the problems with studying very old instruments is that very few original examples survive. Fortunately, they are often depicted in contemporary artists. For example, Fig. 1.5 shows Caravaggio's The Lute Player. The young woman is shown playing a six course lute by plucking it with her fingertips.

Fig. 1.5 The Lute Player by Michelangelo Merisi da Caravaggio (Wikimedia Commons, image is in the public domain)

1.2   The Renaissance and the First Guitars

The 1500s saw the development of the first true guitars. They originally had four courses of two strings each, evolving later to five courses. The Renaissance guitar had a long, thin neck, a slight waist, a flat back and flat top. String tension was varied using tapered pegs as are now found in violins. Figure 1.6 shows a four course guitar from the title page of a collection of guitar compositions by Guillaume Morlaye. It has seven strings in four courses, with the highest pitch string being single.

Only two guitars are presently known to exist from the 16th century [10]; one is by the Portuguese maker Belchior Dias, working in Lisbon. It is dated 1581 and is now in the Royal College of Music in London. The maker of the other instrument is unknown, but it appears to date from about 1590 and may also be by Dias.

The 1581 instrument has a string length of 554 mm (21.81 in). By comparison, modern guitars have string lengths (also called scale lengths) in the range of 629 mm – 648 mm (24.75 in – 25.5 in). It is strung with five courses of paired strings and has a deeply fluted, curved back. The 1590 instrument is larger, with a scale length of about 700 mm (27.56 in). It has a flat back made of eight flat strips of wood joined together with interspersed light strips.

Fig. 1.6 A Four Course Guitar Shown on the Title Page of a Collection of Guitar Music by Guillaume Morlaye, 1552 (Wikimedia Commons, image is in the public domain)

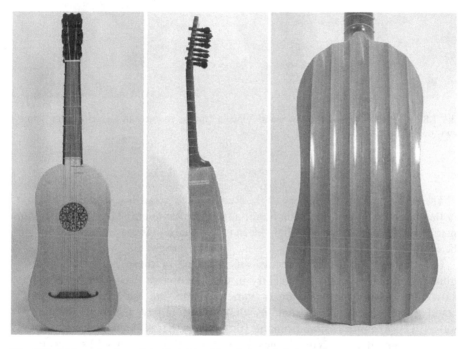

Fig. 1.7 A Reproduction of the Belchior Dias Guitar made by Daniel Larson (Image courtesy Daniel Larson, http://www.daniellarson.com)

Figure 1.7 shows several views of a reproduction Renaissance guitar patterned after the 1581 Dias. Note the original is fitted with five courses of strings while the reproduction has four courses. Notable features include the curved, fluted back, tied-on frets and a fretboard level with the top of the instrument (the soundboard). Note also that the soundhole is covered by a decorative carving similar to those found on lutes.

Fig. 1.8 A Reproduction of the Chambure Vihuela (Image courtesy of Daniel Larson, http://www.daniellarson.com)

The task of tracing the history of the Renaissance guitar is somewhat confused by the existence of the vihuela in Spain during the same era [11]. This, too, is an instrument with a flat top and a slight waist [12]. Indeed, it is easy to mistake one for the other.

Only a few of these instruments survive. One of these is the 'Chambure' vihuela discovered relatively recently in a collection in Paris. The instruments shown in surviving artwork are of varied design and the Chambure doesn't exactly match iconographic examples which appear to have flat backs and narrow bodies. This example, dating from the 16th century, has a fluted back and a wider body than shown in surviving images. Like many vihuelas, it has 12 strings in six courses. Daniel Larson has made a replica based on a study of the original instrument (Fig. 1.8 and 1.9). Another, the 'Guadalupe' is of more conventional design [13].

In Italy and Portugal, a similar instrument was called the viola di mano. The modifier indicates that it was played with fingertips. It was sometimes played with a bow and, in such form, it evolved into the viol (also called the viola da gamba).

While the vihuela is structurally similar to the early guitars, it is musically separate. It was tuned differently and had a separate, distinct repertory. Musically, it may have been more closely related to the lute.

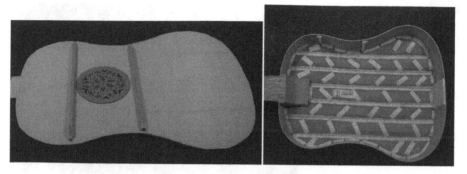

Fig. 1.9 Interior of the Replica Chambure Vihuela (Image courtesy of Daniel Larson, http://www.daniellarson.com)

1.3   The Baroque Era

The Baroque Era is assumed to have lasted roughly from 1600 to 1750 [14]. The Baroque guitar was smaller than modern guitars and more lightly built. Typically, it had five courses of strings, of which either four or five courses were doubled [15]. Thus, both nine and ten stringed instruments appear to have been common. In keeping with the aesthetics of the period, Baroque guitars appear to have been ornate, especially when compared with modern instruments.

The Baroque guitar was a common enough instrument to have been shown in significant works of art. Figure 1.10 shows a painting by Jan Vermeer called The Guitar Player and dates from approximately 1670. The woman in the painting is shown playing a five course instrument with a single soundhole fitted with a decorative rose. Like earlier instruments, the soundboard is level with the fretboard and extends partway up the neck. Both the body and neck are bound with a bold alternating pattern. The bridge is placed well back on the soundboard and doesn't appear to have a separate saddle.

The most famous luthier of the Baroque period, Antonio Stradivari, is known to have made at least two guitars [15, 16]. One, The Rawlins, was made in 1700. It has a flat softwood top with figured maple back and sides, just as his violins do. There is an angled headstock with friction pegs for the five courses of strings. Like earlier instruments, it has a single piece bridge with tied on strings and no separate saddle. The single center soundhole is fitted with an elaborately carved rose.

1.4   Classical and Early Romantic Periods

The Classical period of western music is generally considered to have run from approximately 1750 – 1825. The Romantic period overlaps the Classical period somewhat and is considered to run from about 1800 to 1900. Guitars from the

Fig. 1.10 The Guitar Player
by Jan Vermeer, ca 1670-72
(Wikimedia Commons,
image is in the public domain)

Classical and early Romantic periods show a clear pathway of development. Instruments from the beginning of this era look to us archaic: they have paired courses of gut strings; they are tuned with pegs; and they have fret boards that are level with the sound boards. Some are also very ornate, with elaborate inlays and decorations in the sound holes.

By the early Romantic period, guitars look much more familiar [17]. At first glance, they seem like smaller, thinner versions of modern classical guitars. During this period, the guitar evolved to an instrument with six individual strings rather than pairs of strings. Other refinements during this period included the initial use of fixed metal frets and raising the fret board above the soundboard. Bodies became wider and had more pronounced waists, giving a shape closer to modern 'hourglass' shape. It was also common for instruments of the period to be tuned with mechanical tuning machines rather than friction pegs [18].

Other significant developments are not visible from the outside. Perhaps the most important of these is the use of fan bracing for the soundboard. It had been common to brace the soundboard with a few lateral bars, like those shown in Fig. 1.11. This arrangement is often called ladder bracing. Note that the guitar on the right has one slanted bar. This instrument has also been heavily repaired; the lighter diamond-shaped and rectangular patches are wood that has been glued in to repair cracks.

Fig. 1.12 shows another instrument from the early Romantic Period made by Jean-Nicolas Grobert around 1830. This instrument has a particularly interesting history and

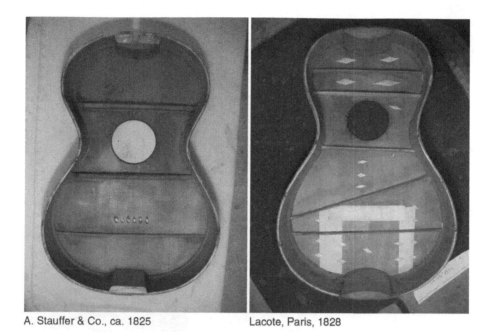

A. Stauffer & Co., ca. 1825   Lacote, Paris, 1828

Fig. 1.11 Guitar Bodies from the Early Romantic Period Showing Ladder Bracing (Images courtesy of Bernhard Kresse, http://www.kresse-gitarren.de)

bears the signatures of Paganini and Berlioz. The guitar was lent to Paganini by Jean-Baptiste Vuillaume in 1838 and later given by Vuillaume to Berlioz [19].

There are several characteristic design features clearly visible in this instrument. Starting at the headstock, it uses friction pegs to tune the strings. Moving down the neck, the fretboard and the soundboard join at the 9th fret, well up on the neck. The neck actually joins the body at the 12th fret. This feature is only possible because the fretboard and soundboard are level with one another.

Since there is no fretboard over the body and the designer wanted to provide more than 12 frets for the player, additional frets have been set directly into the soundboard. The body has a very pronounced waist. Indeed the width of the instrument at the waist is only twice the soundhole diameter. That said, the soundhole is very familiar to a modern viewer. There is only one large soundhole encircled by a decorative ring, the rosette.

The bridge is only slightly wider than the string spread and has characteristic 'whiskers' extending from it. Early bridges were often just simple blocks with holes or slots for the strings. Figure 1.13 shows the bridge on a reproduction baroque guitar. The bridge is a simple block with holes for nine strings in five courses. The end of the string is formed essentially by the block and the loop of string wrapped over it. This would seem to form an imprecise boundary condition. It is, thus, worth noting that the Grobert instrument has a separate saddle that forms a clearly defined

Fig. 1.12 A Guitar by Jean-
Nicolas Grobert, ca. 1830
(Wikimedia Commons,
image is in the public domain)

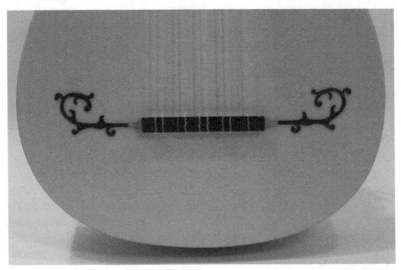

Fig. 1.13 Bridge on a Reproduction Baroque Guitar (Image courtesy of Daniel Larson, http://
www.daniellarson.com)

Fig. 1.14   A Reproduction of a Panormo Guitar, c. 1840 (Image courtesy of Gary Demos, http://www.garydemos.com)

end of the vibrating portion of the string. The part of the bridge that anchors the string so it can be put under tension by the tuner is behind the saddle.

Three of the most important builders of the Romantic period were José Pagés (Cadiz), Louis (or Lewis) Panormo (London) and René Françios Lacôte (Paris). Figure 1.14 shows a replica of an instrument by Panormo (c.1840) by Gary Demos. This instrument has a much narrower body than is now common, but is otherwise familiar. An important feature of this instrument is the fan-braced top. While it doesn't exactly match those that came after it, particularly the one used by Torres, this bracing pattern is an important step between the ladder bracing of earlier instruments and those that followed [20]. Note also that this instrument has a separate, raised fretboard that spans both the neck and the soundboard up to the soundhole.

It was during this period that virtuoso performers achieved international success and compositions for the guitar were published and widely distributed. Probably the most well known of this group was Fernando Sor. In addition to being a very successful performer, he was a composer and left a body of work that is still being performed.

1.5   The Modern Classical Guitar

Antonio de Torres Jurado (usually called Antonio Torres) is perhaps the most well known classical guitar maker and the modern classical guitar essentially began with him [21]. It is not correct to assume that Torres was personally responsible for all the features of his guitars. He was, however, the person who brought all the features together in one instrument. His 1856 instrument is considered one of the most

Fig. 1.15 Replica of an 1856
Torres Made by Kenny Hill
(Image courtesy of Kenny
Hill, http://www.hillguitar.
com)

important designs and one that brought together many of the elements that form the
modern classical guitar. By the 1860s, the form of the modern classical guitar was
essentially in place.

The Torres guitar featured a widened body, fan bracing, geared tuners and a
fretboard raised above the soundboard. The bridge had a rectangular shape, in contrast
to earlier designs that had decorative elements. It had a tie block and a separate saddle
so that string height could be adjusted more easily. The result was a responsive
instrument with enough volume to be suitable for playing to large audiences.

The typical modern classical guitar is very similar to the ones made by Torres in
the late 19th century. Indeed, many luthiers make faithful copies of instruments by
Torres and other well-known builders. Figure 1.15 shows a replica of an 1856
Torres made by Kenny Hill.

It would be quite incorrect to think of the development of the classical guitar as a
unified march toward the modern form of the instrument. Many luthiers, who were
after all, mostly small businessmen with a strong incentive to meet the demands of
the market, experimented with alternative designs. An extremely interesting example is a very rare instrument made by Torres in 1863. It is one of two that he made,
apparently as an experiment in developing an inexpensive instrument for a wider
market [22]. The only known example was found in an advanced state of deteriora-
tion and restored by Richard Bruné.

Fig. 1.16 Interior of a Small
Guitar Made by Torres
in 1863 (Image courtesy
of Richard Bruné, http://
www.rebrune.com)

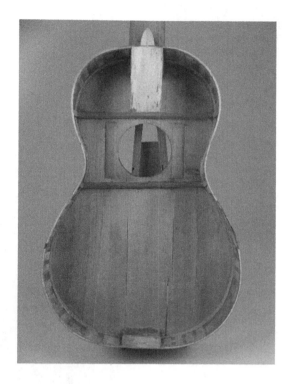

The instrument is small, what might now perhaps be considered a parlor guitar, and has no bracing on the soundboard except for those flanking the soundhole (Fig. 1.16). This seems to suggest that Torres viewed soundboard bracing as a means of tailoring the sound of the instrument rather than a structural element necessary for bearing the load from the strings. Figure 1.17 shows the restored instrument.

There are many slight variations on fan bracing, so it's impossible to point to one example as being the standard. Figure 1.18 shows a seven brace pattern from an instrument made in 1930 by Hermann Hauser. The seven braces are arranged in a fan pattern with the center brace on the joint between the two halves of the soundboard. The small rectangular pieces are reinforcements that have been applied to repair cracks in the soundboard.

Perhaps the most well-known classical guitar of the 20th century was made by Hermann Hauser in 1937 and owned by Andres Segovia [23]. He famously called it the 'greatest guitar of our epoch'. It resides now in the Metropolitan Museum of Art in New York and has been extensively studied. Indeed, some very accurate construction drawings of this instrument are now available (http://www.luth.org and http://www.rebrune.com) and it is not uncommon for luthiers to offer replicas of this instrument. Figure 1.19 shows a replica made by Kenny Hill.

There are certainly luthiers working to refine the classical guitar and who are introducing new design features. Figure 1.20 shows an instrument made in 1977 by R.E. Bruné and purchased in 1984 by Andres Segovia. This instrument has a proprietary bracing pattern developed after both extensive study of older instruments and much building experience.

Fig. 1.17 Restored 1863
Torres (Image courtesy of
Richard Bruné, http://www.
rebrune.com)

There has been much experimentation in alternative bracing schemes for classical
guitars. While some form of traditional fan bracing is still probably the most
common approach, many alternatives have been used. These will be discussed in
more detail later, but a short preview fits here.

One of the most well known alternative bracing schemes was developed by
Michael Kasha and first implemented by Richard Schneider [24, 25]. The Kasha
design uses a highly asymmetric bracing pattern intended to make different portions
of the soundboard have different resonant frequencies. Figure 1.21 shows one
configuration from a patent granted to Kasha in 1978. Note the transverse brace
(part 48) that separates two groups of braces and the outline of the asymmetric
bridge (part 50). The asymmetry in Kasha designs sometimes also extends to the
soundhole location. Some instruments have it placed off center in the upper bout.

Another approach to bracing the tops is to distribute the stiffness across a large
area. Two of the approaches being explored now are lattice bracing and sandwich
structures. Lattice bracing uses a large number of small braces in a crossed
pattern. Typically, the two courses of braces have an included angle of about 60
deg, though many different arrangements have been explored. Figure 1.22 shows
a lattice braced top with conventional cross bars above and below the soundhole.

Fig. 1.18 Interior View of
a 1930 Hauser Showing a Fan
Bracing Pattern (Image
courtesy of Richard Bruné,
http://www.rebrune.com)

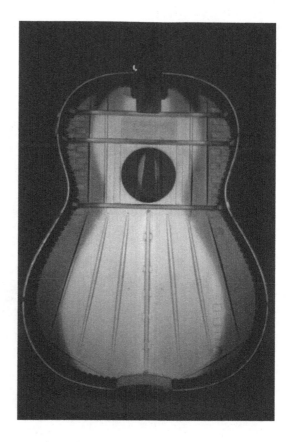

Lattice braces are generally made of wood, though they are sometimes reinforced
with a stiffer material such as unidirectional graphite.

Two other design variations sometimes being used now are elevated fretboards
and side sound ports. Rather than making the top of the neck level with the
soundboard and gluing the fretboard across both, elevated fretboards have an
additional spacer to raise the fretted surface as much as ½" (12.7 mm) from the
top of the fretboard. The object is to improve access to the higher frets.

Side sound ports are intended to change the sound field produced by the
instrument, perhaps directing more of the sound energy to the player. The
soundhole definitely radiates sound from the instrument, but it is usually directed
toward the audience rather than the player. Figure 1.23 shows an experimental
instrument with many holes in the side that could be plugged or opened. The object
of making this instrument was to evaluate a variety of different side port
geometries.

Side ports are not especially common at this writing and there is no consensus on
placement or geometry. Often, they are oval and placed at the upper bout, facing the
player. Figure 1.24 shows a classical guitar with both a side soundhole and an
elevated fretboard.

Fig. 1.19 Replica of the
1937 Hauser Made by Kenny
Hill (Image courtesy of
Kenny Hill, http://www.
hillguitar.com)

Finally, there is the question of materials being used in new classical guitar designs. The classical guitar market, as one might expect, favors tradition. Builders still use solid wood (as opposed to plywood or man-made materials) almost exclusively, though the range of species being used is now much broader than it has been. For now, plywood is used only for the most inexpensive student instruments and man-made materials are hardly used at all for classical guitars. This may well change if high grade wood becomes prohibitively expensive.

1.6   The Modern Steel String Acoustic Guitar

Before the late 1800s, the story of the guitar was essentially a European one. After then, though, the development of the guitar moved largely to the United States, where one of the distinctive changes was the use of steel strings. Wire strings for fretted instrument had been in use for some time [26], but it was only around 1900 that steel strings became widely available. Steel strings offered the possibility of

Fig. 1.20 Classical Guitar
by Richard Brune, Owned
and Played by Andres
Segovia (Image courtesy
of Richard Bruné, http://
www.rebrune.com)

Fig. 1.21 Patent Drawing
of a Kasha Bracing Pattern

March 21, 1978   Sheet 1 to 4   4,079,654

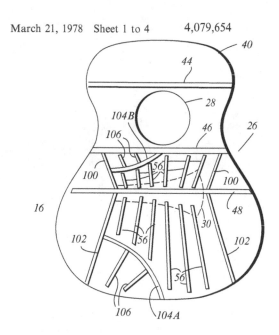

Fig. 1.22 A Lattice-Braced Top by Chris Pantazelos (Image courtesy of Chris Pantazelos, originally published in American Lutherie, http://www.luth.org)

Fig. 1.23 An Experimental Guitar Made to Test the Effect of Side Ports (Image courtesy of R.M. Mottola, http://www.liutaiomottola.com, instrument made by Alan Carruth, http://www.alancarruthluthier.com)

Fig. 1.24 A Classical Guitar with Both Elevated Fretboard and Side Port (Image courtesy of Stephan Connor, http://www.connorguitars.com)

making guitars louder, but the structures of typical classical guitars, then strung with gut strings, were not able to withstand the increased tension. Clearly, new designs were necessary.

Many different designs were offered in the early 1900s, but two that became popular quickly are the flat top with a center soundhole and the archtop. The early steel string flat top guitar is most closely associated with Martin and the early steel string archtop is most closely associated with Gibson. Like Ramirez a few decades before, Martin's success was partly due to a bracing pattern [27]. In this case, the pattern was called X bracing and it allowed the top to withstand the tension of steel strings while being flexible enough to vibrate and radiate sound. This bracing pattern is still widely used and variations of it are probably in most of the steel string acoustic guitars manufactured. Figure 1.25 shows the X-bracing pattern on a cutaway Taylor guitar.

Figure 1.26 shows a Martin 000-18 made in 1927. This design was introduced in the early 1900s and is still in production. In spite of its age, it is very familiar to a modern player. Perhaps the only obvious difference between this and a modern instrument is the slotted headstock. Modern instruments, with the exception of classical guitars, usually have a solid headstock with vertical posts. The modern version of the 000-18 uses a solid headstock and a slightly different bridge. Other, less obvious changes have been introduced over the years, but the result is still clearly a 000-18.

Fig. 1.25 X-Bracing
on a Cutaway Taylor
Acoustic Guitar (Image by
the author, reproduced here
courtesy of Taylor Guitars,
http://www.taylorguitars.com)

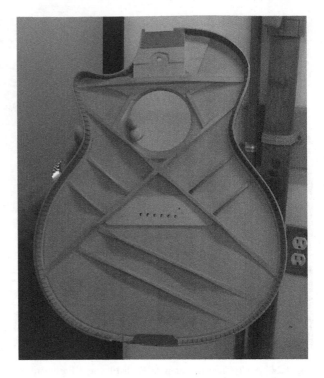

One very popular style of steel stringed acoustic guitar is the Dreadnought. It is a large guitar with a deep body and a squared off body shape. The body is usually attached to the neck at the 14th fret. The original use of the name Dreadnought was due to HMS Dreadnought, a British battleship launched in 1906. Martin manufactured the original Dreadnought body guitar for another company starting in 1916. Martin began producing Dreadnought guitars under the Martin name in 1931, designating them by a model number starting with 'D'. Figure 1.27 shows a 1942 Martin D-45.

One development from around the beginning of the 1900s is the archtop guitar. The traditional archtop guitar is a large acoustic instrument with a domed top and back. Generally, there are only two braces, either placed side by side or placed to form a shallow X [28]. Structurally, archtop guitars occupy a midpoint between flat-top acoustic guitars and members of the violin family – the typical archtop guitar is larger than a viola, but smaller than a cello.

One of the most iconic archtop guitars was the Gibson L-5. It was introduced in 1922 and has been in almost continuous production since then. A famous early player was Mother Maybelle Carter, whose 1928 L-5 is now in the Country Music Hall of Fame in Nashville. Figure 1.28 shows a 1925 model. In addition to the obvious arched top and back, there are several other distinctive features.

Rather than a bridge glued directly to the soundboard, the L-5 uses a floating bridge – one that is held in place only by the tension of the strings. The strings are fixed to the body using a tailpiece anchored at the bottom of the lower bout.

Fig. 1.26 A 1927 Martin
000-18 Acoustic Guitar
(Image courtesy of C.F.
Martin Archives., http://
www.martinguitar.com)

Also, the neck is at a distinct angle to the body (around 5° is common for archtop
guitars) and the fretboard is distinctly elevated above soundboard. Again, this
mimics the construction of the violin as shown in Fig. 1.29.

Archtops were originally unamplified, as were all guitars at the time. However,
they have evolved so that most are fitted with electromagnetic pickups. Archtops
are particularly popular among jazz players who tend to prefer a mellow sound that
results from a hollow body and placement of the pickup away from the end of the
string. As a result, modern versions often have a single magnetic pickup at the end
of the fretboard, about where the 24th fret would be. Figure 1.30 shows a Benedetto
Manhattan jazz guitar. While there is certainly no one best luthier, Bob Benedetto is
one of the best and his archtop guitars are very highly regarded.

Like other types of guitars, steel string acoustic guitars are evolving and many
luthiers are trying out new ideas. One of the most obvious areas of development is
in on-board electronics. Even inexpensive acoustic guitars are often fitted with
piezoelectric sensors and matching preamps so they can be used with standard
guitar amplifiers. Figure 1.31 shows an onboard pre-amplifier mounted in an
Ovation C44. This model works with a piezoelectric sensor under the saddle and
includes a tuner and tone controls. The battery box is to the left of the preamp.

Fig. 1.27 A 1942 Martin D-45 (Image courtesy of C.F. Martin Archives, http://www. martinguitar.com)

One area of development in steel string acoustic guitars is the increasing use of plywood. While there is sometimes a feeling among buyers that plywood is an inferior material for guitars, there is potentially much to recommend it. When well-made, it is durable, uniform and easy to work with. Because it has different stiffness properties than solid wood, the bracing pattern may need to be modified.

Fig. 1.32 shows a Fender 300CE guitar with both on-board electronics and a laminated top. The outer ply is nicely figured wood, so it has the aesthetic appeal of figured wood without the expense (or environmental impact) of using solid figured wood. The result is an adaptable, durable and reasonably-priced guitar with good sound quality.

Of course, there is no mechanical reason that an acoustic guitar couldn't be made exclusively from man-made materials such as fiber-reinforced plastic, usually referred to generically as composites. Indeed, several companies now produce such instruments almost exclusively and others have introduced composite guitars into their product lines.

The first company to achieve significant commercial success with composites in acoustic guitars was Ovation, an offshoot of Kaman, a helicopter manufacturer.

Fig. 1.28 A 1925 Gibson L-5 Archtop Guitar (Image courtesy of Elderly Instruments, http://www.eldery.com)

Figure 1.33 shows an Ovation guitar with their signature pattern of soundholes. It has made largely of composites, with a rounded back and a laminated graphite top. Because of this, it is relatively insensitive to changes in temperature and humidity.

Composites are simply another building material and there is no reason that an instrument with them needs to look radically different than a traditional instrument. Figure 1.34 shows an instrument from RainSong made exclusively of composites, but outwardly traditional in appearance. It also has an on-board pre-amp with tone controls and a tuner.

Other experiments in the design of steel string acoustic guitars are more subtle. Most guitarists have noticed at some point that the hard edges of a large guitar body can be uncomfortable. One possible solution is shown in an instrument by luthier Linda Manzer (Fig. 1.35). She uses a pronounced taper across the width of the instrument that makes it fit more naturally against the player's body.

Finally, there is a relatively new group of hybrid guitars that occupy a middle ground between true acoustic guitars and solid body electric guitars. They are often modeled after solid body electric guitars, but have hollow bodies with thin soundboards as would be found in an acoustic instrument. They are usually intended to only be played while plugged into amplifiers. Indeed, their unamplified sound is generally not good. Rather, they are intended to offer a range of tonal options when amplified and giving the player much more musical flexibility than a conventional acoustic or conventional electric instrument could offer. Figure 1.36 shows a Fender Stratacoustic, a thin-body acoustic electric instrument modeled on the iconic Stratocaster solid body electric guitar.

Fig. 1.29 A Typical Violin for Comparison with an Archtop Guitar (Wikimedia Commons, image is in the public domain)

Fig. 1.30 A Benedetto Manhattan Archtop Jazz Guitar (Image courtesy of Bob Benedetto, http://www.benedettoguitars.com)

1.7   The Electric Guitar

When it became popular to play guitars as part of a larger group of instruments, it quickly became apparent that more volume was often required. One way to increase the volume of an acoustic guitar is to make it bigger. However, there is a clear limit to how large an instrument can be and still be playable. A lower bout width of 17 in (432 mm) and a body length of about 21 in (533 mm) are near the upper limit for a practical instrument, though a few successful instruments have been larger.

Fig. 1.31 A Small Pre-Amp with Equalizer and Tuner (Wikimedia Commons, image is in the public domain)

Fig. 1.32 An Acoustic Guitar Fitted with On-Board Electronics (Image courtesy of Fender Musical Instrument Corp., http://www.fender.com)

Increased understanding of electromagnetism in the late 19th century and widespread commercial distribution of electricity in the early 20th century offered the possibility of amplifying an acoustic guitar.

There is an ongoing discussion about who developed the first electric guitar and it is beyond the scope of this small book to present a comprehensive survey. Rather, here are some important highlights in the development of the electric guitar.

Fig. 1.33 An Ovation Adamas Guitar Made with Composite Materials (Image courtesy of Ovation Guitars, http://www.ovationguitars.com)

Fig. 1.34 An Acoustic Guitar Made of Graphite (Image courtesy of RainSong Guitars, http://www.rainsong.com)

Perhaps the first guitar designed to use an electromagnetic component was patented in 1890 by a young naval officer named George Breed [29]. On Sept 2, 1890, he was awarded patent 435,679 for a "Method of and Apparatus for Producing Musical Sounds by Electricity". Figure 1.37 shows one of the illustrations from this patent.

Breed's design is not an electric guitar in the modern sense because there is no pickup and sound is still radiated from the body of the guitar. However, the instrument is fitted with a large electromagnet that is used to drive the strings by sending pulsed DC current down each string. The intermittent change in the current flowing through the field of the electromagnet exerts a force on the string, causing it to vibrate. Breed's patent also shows the same idea applied to the strings of a piano.

Breed's design was never produced and is important because it seems to be the first time electricity was applied to a stringed, fretted instrument. Much more historically important are several instruments developed in the first part of the 20th century. There were many early attempts at electrically amplified guitars, but

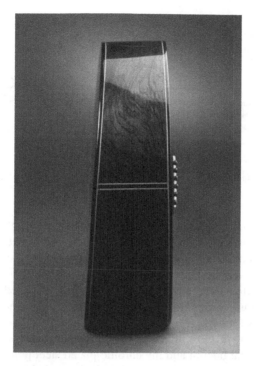

Fig. 1.35 A Tapered Guitar Body on an Acoustic Guitar (Image courtesy of Linda Manzer, http://www.manzer.com)

Fig. 1.36 A Fender Stratacoustic Hybrid Acoustic Electric Guitar (Image courtesy of Fender Musical Instrument Corp., http://www.fender.com)

three are particularly worth mentioning here: the Rickenbacker 'Frying Pan' lap steel guitar, the Gibson ES-150 and the Les Paul 'Log'.

The Rickenbacker 'Frying Pan' was one of the first practical electric guitars and one of the first to use an electromagnetic pickup. It was introduced in response to the popularity of Hawaiian music in the US in the 1920s and 1930s. Figure 1.38 shows this instrument from a patent awarded to George Beauchamp (of Rickenbacker) in 1937.

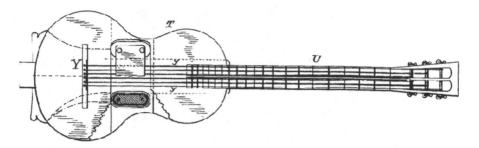

Fig. 1.37 Drawing of George Breed's Electric Guitar from Patent 435,679, issued, Sept 2, 1890

One of the first commercially successful electric guitars was the Gibson ES-150, fitted with a single large electromagnetic pickup. The instrument was introduced in 1936 and became popular with jazz players. While he was not the only player to use this instrument, or even the first, Charlie Christian was the most important. In his short career (he died in 1941 at age 25) he became so associated with ES-150 and the large pickup around which it was made, the pickup was named after him and is even now known as a Charlie Christian.

Figure 1.39 shows a Gibson ES-150 with the distinctive Charlie Christian pickup. The hexagonal bobbin and the chrome plated steel pole piece at its center protrude through the top. The three screws between the bobbin and the bridge are what hold the pickup to the top. The magnets are 13 cm (5 inches) long and the pickup weighs about 900gm (about 2 lb), so it is much larger than a modern pickup.

A guitar that has now become iconic was introduced by Fender in 1949 as the Broadcaster. After some design changes and a name change due to intellectual property concerns, it was re-introduced as the Telecaster. The guitar has been in production ever since. While the ES-150 was an elegant instrument that was relatively expensive to make, the Telecaster was made simply with a slab body and a neck fixed to it with four screws (Fig. 1.40). Also, the electronics pockets are all cut into the front of the instrument for ease of manufacture and assembly. They are covered with a large plastic pickguard and small chromed metal plate that also serves as the mount for the pickup selector switch and the tone and volume controls.

Another iconic solid body guitar is the Gibson Les Paul. It was developed in the early 1950s through a collaboration between Ted McCarty, president of Gibson Guitars, and Les Paul, accomplished guitarist, pop star and inventor. The resulting instrument is a very different one than those produced by Fender. Rather than having a simple slab body with the neck bolted on, the Les Paul has an arched top with the neck glued on at an angle to the body. It has two humbucker pickups with separate tone and volume controls for each. There are many different models of this instrument, ranging from elaborate custom versions to simplified, inexpensive versions intended for beginning guitarists. Figure 1.41 shows a typical model with a figured maple top.

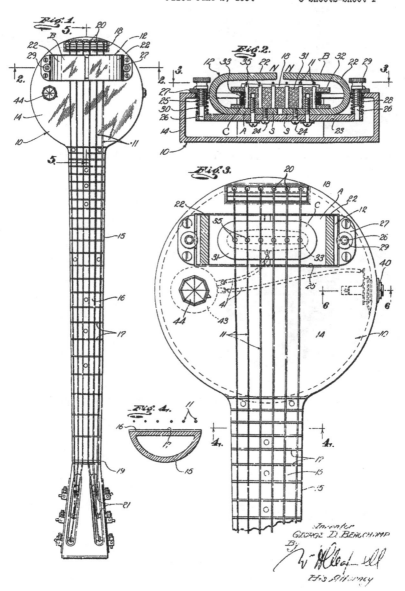

Aug. 10, 1937.   G. D. BEAUCHAMP   2,089,171

ELECTRICAL STRINGED MUSICAL INSTRUMENT

Filed June 2, 1934   3 Sheets—Sheet 1

Fig. 1.38 Patent Drawing of Rickenbacker 'Frying Pan' Lap Steel Guitar

Fig. 1.39 Gibson ES-150 with 'Charlie Christian' Pickup (Wikimedia Commons, image is in the public domain)

Fig. 1.40 1952 Fender Telecaster (Wikimedia Commons, image is in the public domain)

Fig. 1.41 Gibson Les Paul (Wikimedia Commons, image is in the public domain)

Fig. 1.42 – The Eponymous Les Paul Playing One of His Namesake Guitars (Wikimedia Commons, image is in the public domain)

1.8   Future Directions

The world of guitar design and manufacturing is, happily, the center of much creative activity. Luthiers, both professional and amateur, have traditionally embraced tinkering and innovation. There is a willingness among many players and builders to consider even radical designs. One person embodying this tradition is Les Paul (Fig. 1.42). Apart from a long and successful career as a musician, he was responsible for much of the early development of the electric guitar and multi-track recording. Indeed, he was inducted into both the Rock and Roll Hall of Fame (in 1988 in the Early Influence category and again in 2008 as an American Music Master) and the National Inventor's Hall of Fame (in 2005).

While it's impossible to predict the future of the guitar with certainty, some broad directions for innovation seem likely. Some of these are:

• Digital signal processing
• Materials
• Manufacturing processes
• Sound quality research

This is perhaps a dangerous list to present. It may, in following years, seem as myopic as a list of likely innovations in electronics written in 1950 that failed to mention the integrated circuit or the transistor. That said, here is a more detailed explanation.

Digital Signal Processing – Digital electronics are cheap, well-developed and ubiquitous. The same hardware and software technologies that are used for cell phones and digital recorders can be used to modify the electronic signals in guitars. The electric signal from a conventional pickup is analog, but could be converted to a digital signal almost immediately. The digitized signal can be processed on board the guitar or off-board on a dedicated box using a simple, cheap processing chip. When the electrical signal gets to the processor, the sound of the instrument can be modified by software.

Once the sound of the guitar is being modified by software, the only limits in the resulting sound are those imposed the ability to develop processing algorithms. There are, of course, the usual array of delay, reverb, distortion and other familiar effects. However, more fundamental modifications are possible using well-established techniques. For example, a frequency shift filter can be used to change pitch without requiring a change in string tension.

It is conceptually simple to use separate sensors for each string and put a frequency shift sensor in line with each one. Software or electronic hardware could keep the instrument in tune by frequency shifts in the output signals without changes in string tension. Additionally, alternate tunings could be implemented by electronically shifting the frequencies of the appropriate strings. A variety of tunings could be selected instantly with a switch or knob. These concepts have been implemented by Line 6 in their Variax series of instruments. At this writing, a typical Variax has more than 10 alternate tunings that are electronically selectable.

Materials – Advanced materials have been developed for a wide range of engineering applications. It is reasonable to expect some of them to be useful for guitars. For example, fiber reinforced composites, originally developed for other purposes, have been used both for reinforcement and for primary structures in guitars. New forms of composite materials will likely be developed and may include sensors integrated with the materials. Imagine an amplified acoustic guitar in which the structure itself acts as a sensor that could be connected to an amplifier. Rather than motion of the strings or soundboard being sensed at only a few places on the instrument and the resulting signal then sent to the amplifier, the signal could have contributions from the entire structure and the resulting sound might be much richer.

There is also no requirement that new materials be strictly man-made. It is possible that advances in genetic engineering could be applied to create species of trees particularly suited to making instruments. They might be fast growing, have trunks largely free of knots or have particularly desirable mechanical properties. Or, less exotically, it is certainly possible that some form of fiber-reinforced composite could be developed that has acoustic and structural properties similar to the best quality wood.

Manufacturing Processes – Advanced machine tools originally developed for other purposes have been widely applied to guitar manufacturing. Indeed, even the smallest luthier shops can afford computer-controlled routers. Another possibility might be molding; it is, in principle, possible to make a mold whose interior is the shape of a guitar and simply inject some liquid into it that would harden to form a the instrument. This hasn't been done yet, but advances could certainly make it possible.

Alternately, there are rapid prototyping machines, sometimes called 3-D printers, that can produce even very complex parts directly from a geometric model in a computer program – they essentially 'print' a copy of the model. Currently, these machines work mostly with plastics that are too weak for stringed instruments. 3-D printing in metal is now possible, but very expensive. However, it may not be unreasonable to think that further refinements would allow a designer to simply 'print out' a working guitar when the geometric model was complete.

Sound Quality Research – This may seem like an arcane thing to list here, but there is great potential. There is currently no analytical model that accurately describes or predicts the tonal quality of guitars. Probably, this is the part of instrument design that we know least about and there is plenty of room for advancement. It is conceivable that an instrument could be designed using computer software that would accurately compute the resulting sound. It might even be possible to specify desired tonal characteristics and then develop an instrument that produced the desired sound. At this writing, though, such an advancement seems to be well in the future.

1.9   The Guitar Business

The design and manufacture of guitars is, of course, strongly conditioned by business pressures. No overview would be complete without a brief discussion of the nature of the guitar business. There are many different ways of trying to impose some sort of organization to the larger picture. One is to distinguish by the size of the manufacturers.

The largest manufacturers produce hundreds or even thousands of instruments per day. At this writing the largest manufacturing operations are in Asia because of lower costs. However, there are numerous guitar factories in North America, some producing hundreds of instruments per work day. The size of production facilities runs down through those manufacturing a few or a few dozen instruments per day to individual builders who might make only a few instruments per year. For the purposes of this brief introduction, we can simply distinguish between small batch manufacturers and large scale production.

1.9.1   Small Batch Manufacturers

One of the most appealing features of the guitar is that a determined person can make one with very few tools and basic, readily available, materials. Garages and basements all over the world are home to people making instruments, either for their own satisfaction or as a way of supporting themselves. Indeed, the author has made many guitars over the last 20 years in a garage and two small basement workshops.

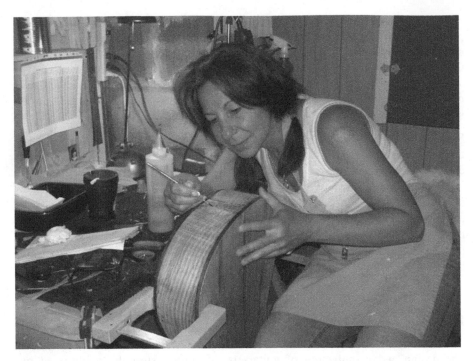

Fig. 1.43 Sophie Karolidis Working on A Classical Guitar in Her Home Shop (Image courtesy of Sophie Karolidis)

An individual luthier might make anywhere from a few instruments per year to a few instruments per month; typical output for a professional luthier might be 24 instruments per year. Output depends on many things including skill level, complexity of the instruments and available tools. Instruments are routinely made using only hand tools and some builders prefer the craft experience of using primarily hand methods. However, most luthiers have access to a range of power tools and some now use computer-controlled tools (called CNC for Computer Numerically Controlled) such as CNC routers. It seems likely that CNC tools will become more common as they get progressively cheaper.

Many working luthiers find it attractive to start with a simple space and set of tools. The initial investment can be very modest and, with some skill and patience, the results can be quite good. Figure 1.43 shows Canadian luthier Sophie Karolidis working at one of three workstations in her home shop. Her total working area is about 300 square feet (27.9 square meters).

Individual luthiers often have a business strategy that fits their own situation and their own preferences. Karolidis has chosen to offer a classic design at an accessible price. Her instruments are built from plans for a 1933 Santos Hernandez instrument drawn by Roy Courtnall (http://www.guitarplans.co.uk). Figure 1.44 shows her Loba Blanca model. This instrument has an Englemann spruce top and Alaskan yellow cedar back and sides.

Fig. 1.44 Blanca Loba Model Classical Guitar (Image courtesy Sophie Karolidis)

Karolidis is also representative of many independent luthiers in that she is largely self taught. She has learned her craft from books and videos and was also fortunate to spend time with a local flamenco guitar maker who gave her some informal instruction.

Kari Hahn is a young luthier whose story is representative of many others starting in the field. Her woodworking career began in Nome, Alaska at age six when she started making toys in her father's workshop. She is now in Portland, Oregon, where she makes both marimbas and guitars. Figure 1.45 shows Kari at work in her small, tidy shop.

Kari's work hints at a love of the outdoors. Figure 1.46 shows a sound hole decoration in the form of a sunflower. The petals are made of yellow heart and apricot wood. Note that the inlay also covers the end of the fretboard.

While many, probably most, luthiers are still self-taught, there are a number of formal training programs in the United States and Europe. It is common for individual luthiers to start work after formal training and to support themselves with a combination of building and repairs. One of these is Austin McKee, a young luthier working in Ottowa, Indiana (USA). At this writing, he has made more than 30 instruments and supplements this business with a steady stream of repairs. Figure 1.47 shows his workshop.

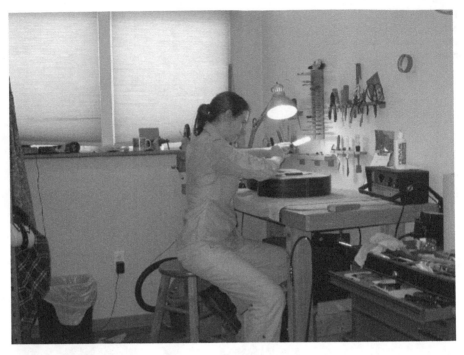

Fig. 1.45 Luthier Kari Hahn at Work (Image courtesy of Kari Hahn, http://www.klathguitars. com)

Fig. 1.46 A Sunflower Soundhole Decoration by Kari Hahn (Image courtesy of Kari Hahn, http:// www.klathguitars.com)

Fig. 1.47 Workshop of Luthier Austin McKee (Image courtesy of Austin McKee, http://www. austinmckee.com)

One of the most successful individual luthiers currently working is Richard Bruné (http://www.rebrune.com), who works in Evanston, Illinois, in the northern part of Chicago. He began making guitars in his house in Dayton, Ohio in 1966. He became a skilled flamenco player, playing in both the US and in Mexico until 1972. He set up shop in Evanston in 1973 and has worked there since. He now makes and restores very fine instruments. He has the rare distinction of having had Andres Segovia, the greatest classical guitarist of the 20th century, as a customer. Segovia had a reputation for being an exceptionally demanding guitar buyer who would play only the finest instruments.

After a significant upgrade in 1995, Bruné's shop includes both work areas and a showroom lined with rosewood cabinets full of guitars. The whole space is climate controlled to ensure that wood doesn't change shape or generate internal stresses with changes in temperature and humidity. Bruné uses classical methods when making and restoring instruments, even to the extent of using hot hide glue rather than polymer-based wood glue. Figure 1.48 shows a 1952 Hauser guitar being restored on his work bench.

Since the cost of basic CNC machines has decreased to the point that they are affordable for individuals, they are starting to appear in the shops of individual luthiers. There are two ways of using CNC tools in a small shop. The first is to simply use them to make parts directly and the second is to use CNC tools to make

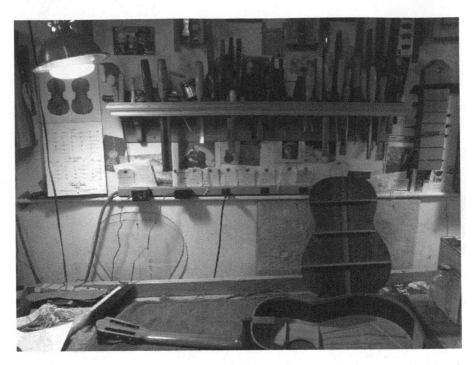

Fig. 1.48 Workbench of Richard Bruné (Image courtesy of R.E. Bruné, http://www.rebrune.com)

accurate templates. Figure 1.49 shows Josh Hurst, an accomplished designer and builder, manufacturing necks with a small CNC router.

There are generally two steps in preparing to mill a part using a CNC tool. The first is to make a geometric model of the part using computer aided design (CAD) software. The resulting model could be a true 3-D model (often called a solid model), as shown in Fig. 1.50, or a 2-D model that defines the outlines and the shape of the pockets for the electronics, neck, pickups, etc. There are many commercial CAD packages on the market and some open source packages as well. The set of features required to model a guitar depends on the design. A solid body electric guitar with a slab-shaped body (like the Telecaster) requires only basic modeling capabilities. However, a more complex shape like an archtop guitar might require more sophisticated features.

Once the mathematical model of the part is complete, the next step is to create a set of instructions that tell the milling machine how to make the part. These instructions are called tool paths and literally tell the machine what paths the cutting tool should take through space. For all but the simplest parts, the tool paths are far too complex to be defined manually. Rather, they are calculated using computer aided manufacturing (CAM) software.

Finally, once the tool paths have been created, the parts can be machined. For solid electric guitar bodies, it is typical to make a rectangular body blank from

Fig. 1.49 A Small CNC Router Being Set Up to Make Guitar Necks (Image courtesy of Josh Hurst)

Figure 1.50 CAD Model of an Experimental Arched Top with Integral Braces (Model by Eddy Efendy)

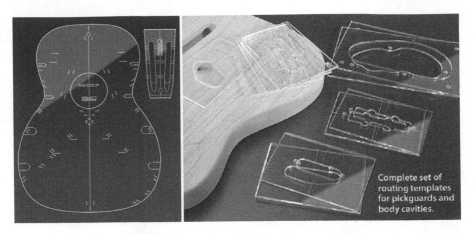

Complete set of
routing templates
for pickguards and
body cavities.

Fig. 1.51 Representative Templates Available for Luthiers (Images courtesy of Stewart MacDonald, http://www.stewmac.com)

which the body will be milled. The body blank is fixed to the cutting table, often with the aid of vacuum, and the tool path commands are run.

The second way of applying CNC tools is to use them to make jigs and templates. When budgets don't allow purchasing CNC tools, having a selection of precise templates allows for more accurate and repeatable work. While templates can certainly be manufactured on a custom basis, some luthiery suppliers sell commonly needed templates, such as those for pickups. Figure 1.51 shows some commercially available templates.

1.9.2   Large Scale Production

Clearly, most instruments are made in factories designed to produce similar instruments in large numbers. Different companies have differing manufacturing capacities; small companies might produce a few dozen instruments per day while a large one might produce hundreds per day. Taylor Guitars in El Cajon, California is an example of a large manufacturer that uses advanced manufacturing methods to produce guitars.

In 1974, Bob Taylor co-founded, along with Kurt Listug and Steve Schemmer, what eventually became Taylor Guitars [30]. It started as a fairly typical small guitar manufacturing company, but eventually evolved into a large manufacturer with an international market and some of the most advanced processes in the industry. The company uses computer-controlled machinery for many production process and computer manufactured equipment and fixtures for nearly everything else.

The result is a production process with very low build variation. It is an axiom of manufacturing that build variation and build quality are inversely related. Industrial

Fig. 1.52 Production Floor at Taylor Guitars in Early 2010 (Image by the author, reproduced courtesy of Taylor Guitars, http://www.taylorguitars.com)

quality improvement practices have a strong focus on repeatability – making everything the same every time. When a product doesn't turn out the same every time, the difference is called build variation [31].

One of the most useful techniques for improving quality is continuous improvement – making small changes in the process often to improve the resulting product. However, if build variation is large, the effect of small changes in the manufacturing process might be lost in the 'noise'. Conversely, if build variation is low, it becomes possible to continuously make small improvements. Over time, the accumulated result can be major improvements in the quality of the product.

Taylor uses CNC equipment, controlled temperature and humidity, well-designed fixtures, carefully conditioned wood and many other techniques to manufacture instruments very precisely. Figure 1.52 shows part of the production floor at Taylor as it appeared in early 2010. The space is clean, orderly and is clearly dominated by CNC equipment.

In a production sequence as complicated as that for a guitar, it is important to account for as many sources of variation as possible. For example, wood is very sensitive to changes in humidity; it expands when the humidity increases and shrinks when the humidity decreases. In addition, wood can have internal stresses generated by the process of felling the tree, cutting it into planks and drying them.

Fig. 1.53 A Pallet of Rosewood for Acoustic Guitar Backs Being Acclimated (Image by the author, reproduced here courtesy of Taylor Guitars, http://www.taylorguitars.com)

To limit changes in the behavior of the material in the production process, most luthiers acclimate wood to the environment where the instrument will be made. Taylor stacks the thin plates used for tops and backs, interleaving them with specially designed plastic spacers called stickers. Figure 1.53 shows a pallet of rosewood backs separated by stickers.

Taylor's production process has been shown to be very repeatable [32]. This is most likely due to their ability to work to very tight tolerances. Because they can control the dimensions of their parts so precisely, they are able to use building methods that wouldn't be possible otherwise. Figure 1.54 shows a CNC laser at Taylor being used to cut tops from thin plates of solid spruce.

The laser cuts cleanly and is also used to put alignment marks on parts. Figure 1.55 shows an alignment mark on the inside of a top that will eventually become part of a T5.

Another feature made possible by precise manufacturing methods is Taylor's neck attachment method. Bolted-on necks are increasingly common among acoustic guitars, but this design includes pockets for two spacers that ensure precise neck alignment (Fig. 1.56). During final assembly, an array of precisely-manufactured spacers is available and the correct thicknesses are chosen to make sure the neck is correctly aligned with the body. Since the neck is easy to remove, the spacers can also be replaced if the instrument deforms slightly as it ages.

Wood has been the overwhelmingly preferred material for guitars, but composites have also been used. A small manufacturer making instruments

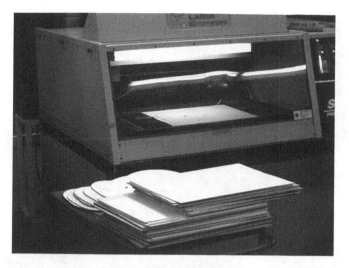

Fig. 1.54 Tops Being Cut Using a CNC Laser (Image by the author, reproduced here, courtesy Taylor of Guitars, http://www.taylorguitars.com)

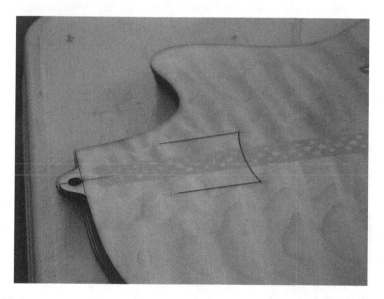

Fig. 1.55 Laser-Etched Assembly Marks on the Interior of a T5 Top (Image by the author, reproduced here, courtesy of Taylor Guitars, http://www.taylorguitars.com)

exclusively from graphite is RainSong Guitars in Woodlinville, Washington (USA). The build process for these instruments is very different from that for a typical wood instrument.

Figure 1.57 shows a part being cut from a roll of pre-impregnated fabric (usually called pre-preg). Pre-preg fabric has the epoxy already applied so that it can be

Fig. 1.56 A Completed Acoustic Guitar Body Showing a Pickup and Pockets for Neck Shims (Image by the author, reproduced here, courtesy of Taylor Guitars, http://www.taylorguitars.com)

Fig. 1.57 A Guitar Back Being Cut from Graphite Fabric (Image courtesy of RainSong Guitars, http://www.rainsong.com)

simply cut to shape, placed in a form and heated to cure the epoxy. Since the material is supplied in rolls, it is easy to store on a rack and then simply cut parts out as needed.

Figure 1.58 shows uncured pre-preg being laid into a mold. The result will be a single part that forms the back and sides of the instrument.

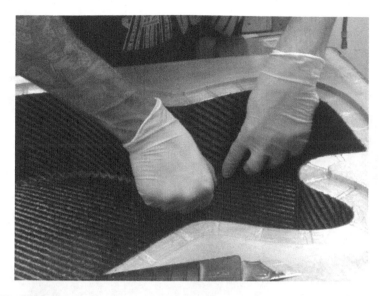

Fig. 1.58 Graphite Fabric Being Laid into a Mold to Form a Single-Piece Back and Sides (Image courtesy of RainSong Guitars, http://www.rainsong.com)

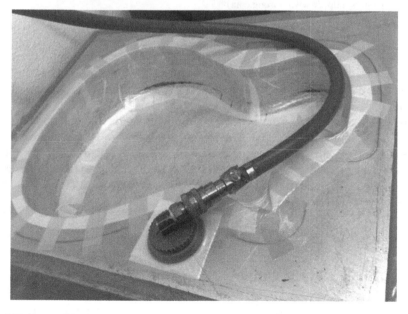

Fig. 1.59 Integral Back and Side Being Cured Under Vacuum (Image courtesy of RainSong Guitars, http://www.rainsong.com)

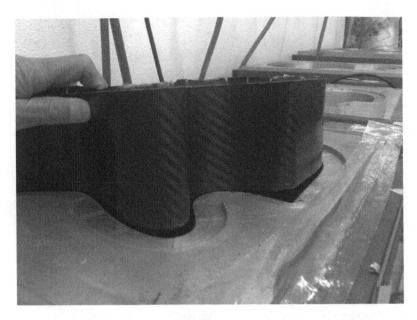

Fig. 1.60 A Cured Graphite Back with Sides Being Removed from the Mold (Image courtesy of RainSong Guitars, http://www.rainsong.com)

With the fabric in the mold, the next step is to apply a vacuum to push the fabric firmly against the inner surface of the mold. Figure 1.59 shows the mold with a light covering material over the black graphite fabric. An airtight layer of plastic allows a vacuum to be drawn through the fitting below the waist.

The epoxy cures when heated and it is common for molds with the pre-preg to be placed in an autoclave that applies heat and pressure. While this approach has proven to work well, autoclaves tend to be large and expensive. RainSong has, instead, developed molds with integral heaters. The result is a process that takes very little space and makes high quality parts. Figure 1.60 shows the cured body part being removed from the mold.

1.10 Timeline of the Guitar

Since the development of the guitar happened in a number of different places and over a long time, it is helpful to have a timeline to orient the various precursors with respect to well-known events.

Development of the Guitar Before 1800

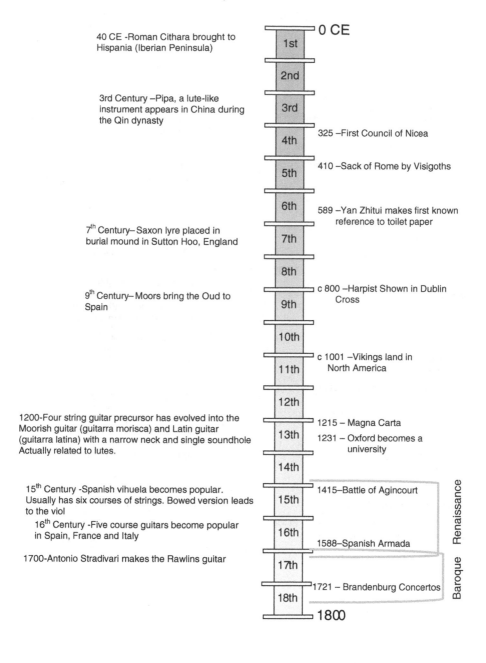

40 CE -Roman Cithara brought to Hispania (Iberian Peninsula)

0 CE

1st

2nd

3rd Century –Pipa, a lute-like instrument appears in China during the Qin dynasty

3rd

4th   325 –First Council of Nicea

5th   410 –Sack of Rome by Visigoths

6th   589 –Yan Zhitui makes first known reference to toilet paper

7th Century– Saxon lyre placed in burial mound in Sutton Hoo, England

7th

8th

9th Century– Moors bring the Oud to Spain

9th   c 800 –Harpist Shown in Dublin Cross

10th

c 1001 –Vikings land in North America

11th

12th

1200-Four string guitar precursor has evolved into the Moorish guitar (guitarra morisca) and Latin guitar (guitarra latina) with a narrow neck and single soundhole Actually related to lutes.

13th   1215 – Magna Carta
1231 – Oxford becomes a university

14th

15th Century -Spanish vihuela becomes popular. Usually has six courses of strings. Bowed version leads to the viol

15th   1415–Battle of Agincourt

16th Century -Five course guitars become popular in Spain, France and Italy

16th
1588–Spanish Armada

1700-Antonio Stradivari makes the Rawlins guitar

17th

18th   1721 – Brandenburg Concertos

1800

Renaissance

Baroque

Development of the Guitar After 1800

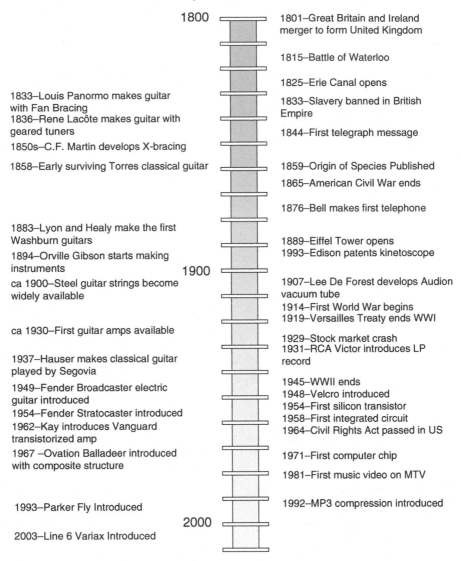

1800

1801–Great Britain and Ireland merger to form United Kingdom

1815–Battle of Waterloo

1825–Erie Canal opens

1833–Louis Panormo makes guitar with Fan Bracing
1836–Rene Lacôte makes guitar with geared tuners

1833–Slavery banned in British Empire

1844–First telegraph message

1850s–C.F. Martin develops X-bracing

1858–Early surviving Torres classical guitar

1859–Origin of Species Published

1865–American Civil War ends

1876–Bell makes first telephone

1883–Lyon and Healy make the first Washburn guitars

1894–Orville Gibson starts making instruments

1889–Eiffel Tower opens
1993–Edison patents kinetoscope

1900

ca 1900–Steel guitar strings become widely available

1907–Lee De Forest develops Audion vacuum tube
1914–First World War begins
1919–Versailles Treaty ends WWI

ca 1930–First guitar amps available

1929–Stock market crash
1931–RCA Victor introduces LP record

1937–Hauser makes classical guitar played by Segovia

1949–Fender Broadcaster electric guitar introduced
1954–Fender Stratocaster introduced
1962–Kay introduces Vanguard transistorized amp
1967 –Ovation Balladeer introduced with composite structure

1945–WWII ends
1948–Velcro introduced
1954–First silicon transistor
1958–First integrated circuit
1964–Civil Rights Act passed in US

1971–First computer chip

1981–First music video on MTV

1993–Parker Fly Introduced

1992–MP3 compression introduced

2000

2003–Line 6 Variax Introduced

References

1. Kilmer AD (1998) The Musical Instruments from Ur and Ancient Mesopotamian Music. Expedition 40(2):12
2. Montagu J (2007) Origins and Development of Musical Instruments. Scarecrow Press
3. Coelho VA, ed. (2003) The Cambridge Companion to the Guitar. Cambridge University Press
4. Grunfeld, FV (1969) The Art and Times of the Guitar – An Illustrated History of Guitars and Guitarists, Macmillan, London
5. Osborne N, ed. (2002) The Classical Guitar Book: A Complete History. Backbeat Books
6. Kasha M (1968) A New Look at the History of the Classic Guitar. Guitar Review 30:3–12
7. Maas M and Snyder JM (1989) Stringed Instruments of Ancient Greece. New Haven: Yale University Press, 1989
8. Tyler J and Sparks P (2007) The Guitar and Its Music: From the Renaissance to the Classical Era. Oxford University Press
9. Smith DA (2002) A History of the Lute from Antiquity to the Renaissance. Lute Society of America
10. Turnbull H (1992) The Guitar From the Renaissance to the Present Day. Bold Strummer
11. Chase, Gilbert (1942) Guitar and Vihuela: A Clarification. Bulletin of the American Musicological Society 6:13–14
12. Weisman M (1982) The Paris Vihuela Reconstructed. The Galpin Society Journal 35:68–77
13. Prynne, Michael (1963) A Surviving Vihuela de Mano. The Galpin Society Journal 16:22–27
14. Oxford Companion to Music. Oxford Music Online. http://www.oxfordmusiconline.com (2010). Accessed 10 Sept 2010
15. Dart, Thurston (1954) Instruments in the Ashmolean Museum. The Galpin Society Journal 7:7–10
16. Faber T (2006) Stradivari's Genius: Five Violins, One Cello and Three Centuries of Enduring Perfection. Random House
17. Usher T (1956) The Spanish Guitar in the 19th and 20th Centuries. The Galpin Society Journal 9:5–36
18. Bruné ME (2005) Classic Instruments: James Ashborn, Innovative Entrepreneur. Vintage Guitar, April
19. Bone PJ (1914) The Guitar and Mandolin: Biographies of Celebrated Players and Composers for These Instruments
20. Martin D (1998) Innovation and the Development of the Modern Six-String Guitar. The Galpin Society Journal 51:86–109
21. Bruné RE (1997) Classic Instruments: Antonio de Torres, The Birth of the Modern Guitar. Vintage Guitar, October
22. Bruné RE (2008) Classic Instruments: Antonio de Torres 1863. Vintage Guitar, May
23. Bruné RE (2003) Classic Instruments: Andre Segovia's Hauser, Built in 1937. Vintage Guitar, October
24. Perlmeter A (1970) Redesigning the Guitar. Science News 98(8/9): 180–181
25. Kunzig R (2000) Roll Over, Segovia. Discover 21(6):35–36
26. Martin D (2006) The Early Wire-Strung Guitar. The Galpin Society Journal 59:123
27. Johnston R and Boak D (2008) Martin Guitars: A History. Hal Leonard
28. Benedetto R (1996) Making and Archtop Guitar. Centerstream Publications
29. Hill M (2008) George Breed and His Electrified Guitar of 1890. The Galpin Society Journal 61:193
30. Simmons MJ (2004) Taylor Guitars: 30 Years of a New American Classic. PPV Medien
31. French M and Efendy E (2010) Reducing Build Variation in a Group of Arched Plates. American Lutherie 101:61–62
32. French M and Brubaker K (2007) Build Variation in a Group of Acoustic Guitars. American Lutherie 90: 28–31

Chapter 2
Basic Physics

The most subtle elements of how guitars produce sound can be complex, but the basic principles are fairly simple. There are basically two different sound producing mechanisms to consider, one for acoustic guitars and one for electric guitars. While the way the sound field is produced differs, both types of instrument start with string vibration. The obvious difference is that acoustic guitars use the resulting vibration of the flexible body to generate sound while electric guitars use pickups to sense string motion and turn it into an electrical signal that can be amplified.

However, before we can understand the basic mechanics of how guitars work, we need a brief introduction to the patterns that describe how music works. The math that describes the relationships between notes also dictates how frets are laid out on the neck.

2.1 Structure of Music

Any piece of music is nothing more than a succession of notes. They have a variety of different frequencies (what musicians call pitches), but they don't have just a random collection of frequencies. Rather, there is a clear and simple mathematical structure that specifies precise frequencies for notes. Because this structure affects many aspects of the guitar's design, it makes sense to sort it out first so we can have a clear framework for later topics.

If there is anything confusing or intimidating about basic music theory, it might be that there are many names to memorize and that it's hard to discern an underlying structure from them. When described in terms of simple mathematical relations, though, the underlying patterns are easier to sort out.

Music is defined in terms of intervals, the frequency ratios of different notes. Musicians are taught to think of intervals as the distance between notes and they are the building blocks for all that follows. The smallest possible interval is variously called a semi-tone, a half tone or a half step. The interval made from two of these

R.M. French, *Technology of the Guitar*, DOI 10.1007/978-1-4614-1921-1_2,
© Springer Science+Business Media New York 2012

Number of Half Steps	Name
1	Half Step
2	Whole Step
3	Minor Third
4	Major Third
5	Perfect Fourth
6	Augmented Fourth/Diminished Fifth
7	Perfect Fifth
8	Minor Sixth
9	Major Sixth
10	Minor Seventh
11	Major Seventh
12	Octave

Table 2.1 Names of Musical Intervals

together is called a whole tone or a whole step. For clarity, these will be called here a half step and a whole step. Musicians have a clear naming system for describing intervals as shown in Table 2.1.

The earliest attempt to describe the mathematical relationships of notes dates back to Pythagoras. The Pythagorean scale uses intervals defined by ratios of small integers. For example, the octave is defined by a frequency ratio of 2:1, a perfect fifth is a frequency ratio of 3:2 and a perfect fourth is a ratio of 4:3. This system is tidy, but creates some practical difficulties. Some combinations of intervals aren't consistent and would seem to require a single note to have different frequencies depending on how it is used.

The solution to this problem is to slightly modify the ideal Pythagorean ratios in a process called temperament. There are many different schemes for temperament – think of Bach's Well-Tempered Klavier – but the most common system now is called equal temperament. Equal tempering uses one number for the frequency ratio describing a half step and builds the rest of the scale from it. That number is approximately 1.0595. While this number might not have made Pythagoras very happy, it works quite well and is easy to derive.

We have already established that every note can be described by the ratio of its frequency to some other one. Let's say that we've picked some note to start with (about which more anon) and we'll call the frequency of that note f_0. We'll call the next note, which is a half step higher in frequency, f_1. For now, let's also call the ratio of these two frequencies r, so $r = f_1 / f_0$. Then, the ratios of all the notes in the previous table can be clearly described.

From Table 2.2, the ratio of the first and 12th frequencies is $f_{12} / f_0 = r^{12}$. The Pythagorean system established an octave as a doubling of the frequency – a frequency ratio of 2:1. This also means that $f_{12} / f_0 = 2$. Finally

$$f_{12}/f_0 = 2 = r^{12} \quad \rightarrow \quad r = 2^{1/12} \approx 1.05946 \quad (2.1)$$

So now we can calculate exactly the relationship between the frequencies of any two notes. However, we need to be able to calculate the actual frequencies of all

Table 2.2 Expressions for Intervals

Number of Half Steps	Relationship
0	f_0
1	$f_1 = r f_0$
2	$f_2 = r f_1 = r^2 f_0$
3	$f_3 = r f_2 = r^3 f_0$
4	$f_4 = r^4 f_0$
5	$f_5 = r^5 f_0$
6	$f_6 = r^6 f_0$
7	$f_7 = r^7 f_0$
8	$f_8 = r^8 f_0$
9	$f_9 = r^9 f_0$
10	$f_{10} = r^{10} f_0$
11	$f_{11} = r^{11} f_0$
12	$f_{12} = r^{12} f_0$

Table 2.3 Names of the Notes

Number	Name
1	A
2	A♯ / B♭
3	B
4	C
5	C♯ / D♭
6	D
7	D♯ / E♭
8	E
9	F
10	F♯ / G♭
11	G
12	G♯ / A♭

notes, not just the relationships between them. In order to do that, we need to know the frequency of just one of the notes. Knowing the exact frequency of any single note means that the frequencies of all notes can be calculated. An international standard [1] has defined the frequency of a note called A_4 to be 440 Hz. This definition is more helpful if it is clear how notes are named.

It is much more convenient for notes to have names rather than just be referred to by their frequencies. Thus all notes have been given letter names from A to G. The sharp, ♯, indicates that a letter note is raised by a half-step and the flat, ♭, indicates that a letter note is lowered by a half step. There are 12 half steps in an octave and every note in an octave has a unique name. There are many more than 12 notes within the human hearing range, so names are repeated. Starting with A, the notes in an octave are shown in Table 2.3.

Some notes have two different names like A♯ / B♭. This is a legacy of older temperaments in which A♯ and B♭ sometimes really were slightly different notes. In the even tempered scale, they have the same frequencies. Also, some scales cannot be written on the music staff using sharps and must be written using flats instead.

Table 2.4 Equal Tempered Notes in the Human Hearing Range (Hz)

E_0	20.602	E_3	164.81	E_6	1318.5	E_9	10548
F_0	21.827	F_3	174.61	F_6	1396.9	F_9	11175
	23.125		185.00		1480.0		11840
G_0	24.500	$\mathbf{G_3}$	**196.00**	G_6	1568.0	G_9	12544
	25.957		207.65		1661.2		13290
A_0	27.500	A_3	220.00	A_6	1760.0	A_9	14080
	29.135		233.08		1864.7		14917
B_0	30.868	$\mathbf{B_3}$	**246.94**	B_6	1975.5	B_9	15804
C_1	32.703	C_4	261.63	C_7	2093.0	C_{10}	16744
	34.648		277.18		2217.5		17740
D_1	36.708	D_4	293.66	D_7	2349.3	D_{10}	18795
	38.891		311.13		2489.0		19912
E_1	41.203	$\mathbf{E_4}$	**329.63**	E_7	2637.0	E_{10}	21096
F_1	43.654	F_4	349.23	F_7	2793.8		
	46.249		369.99		2960.0		
G_1	48.999	G_4	392.00	G_7	3136.0		
	51.913		415.30		3322.4		
A_1	55.000	A_4	440.00	A_7	3520.0		
	58.270		466.16		3729.3		
B_1	61.735	B_4	493.88	B_7	3951.1		
C_2	65.406	C_5	523.25	C_8	4186.0		
	69.296		554.37		4434.9		
D_2	73.416	D_5	587.33	D_8	4698.6		
	77.782		622.25		4978.0		
$\mathbf{E_2}$	**82.407**	E_5	659.26	E_8	5274.0		
F_2	87.307	F_5	698.46	F_8	5587.7		
	92.499		739.99		5919.9		
G_2	97.999	G_5	783.99	G_8	6271.9		
	103.83		830.61		6644.9		
$\mathbf{A_2}$	**110.00**	A_5	880.00	A_8	7040.0		
	116.54		932.33		7458.6		
B_2	123.47	B_5	987.77	B_8	7902.1		
C_3	130.81	C_6	1046.5	C_9	8372.0		
	138.59		1108.7		8869.8		
$\mathbf{D_3}$	**146.83**	D_6	1174.7	D_9	9397.3		
	155.56		1244.5		9956.0		

The human hearing range is generally assumed to be 20 Hz – 20,000 Hz, though age and accumulated exposure to loud noises reduces that range in almost all people (I can't hear anything above about 12 kHz). Since the names of the notes repeat themselves in a regular cycle, there are actually 10 A notes within the nominal human hearing range. The international standard defines the frequency of the fourth A in the human hearing range, A_4, to be 440 Hz. This single definition combined with an accurate value for r allows the frequencies of all the notes in the scale to be calculated.

Table 2.4 shows all the equal tempered notes in the human hearing range. The standard tuning of the six strings of a guitar is EADGBE and these notes are

Major Scale

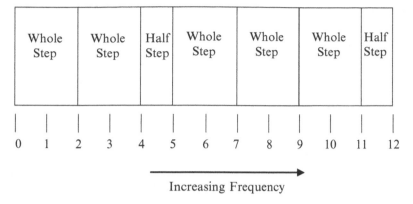

Increasing Frequency

Minor Scale

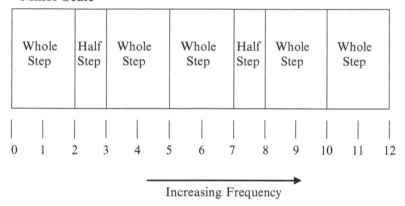

Increasing Frequency

Fig. 2.1 The patterns of whole steps and half steps that form major and minor scales

shown in bold. The lowest frequency a guitar can produce at standard tuning is 82.4 Hz. The lowest note any human could be expected to hear is E_0 at 20.6 Hz.

Before moving on to the physics of the guitar, it is worth mentioning the basis of musical scales. There are 12 half steps in an octave, but the very name suggests that there should only be eight. While there are 12 possible notes in an octave, only eight of them are used for a scale. These eight notes are defined by a set pattern of intervals. For a major scale, the interval is 1 1 ½ 1 1 1 ½. There are other scales with other interval patterns, but the two most common are the major and minor scales. The interval patterns for these two scales are shown in Fig. 2.1.

To define a scale, it is only necessary to pick a starting note and step through the list of possible notes using the interval pattern. The name of the scale is determined by the name of the starting note and the interval pattern used. For example, the C major and A major scales are shown in Table 2.5. There are patterns for all the

Number	Interval	C Major	A Major
Table 2.5 C Major and A Major Scales			
1	Start	C	A
		C♯ / D♭	A♯ / B♭
2	Whole	D	B
		D♯ / E♭	C
3	Whole	E	C♯ / D♭
4	Half	F	D
		F♯ / G♭	D♯ / E♭
5	Whole	G	E
		G♯ / A♭	F
6	Whole	A	F♯ / G♭
		A♯ / B♭	G
7	Whole	B	G♯ / A♭
8	Half	C	A

different scales, but major and minor are by far the most common ones and enough to make the point here.

In a similar fashion, chords are defined as three or more notes with specific frequency ratios. That means they also be defined as three specific notes in a scale. For example, a major chord is made from notes 1, 3 and 5 in a major scale. Thus, a C major chord contains the notes C, E and G. An A major chord contains the notes A, $C^{\#}$ and E. There are similar 'recipes' for all chords.

2.2   Acoustics

We all know that sound is formed from pressure waves propagating through the air. To understand how musical instruments work, it helps to know something about how those pressure waves are formed, how they propagate and how we perceive them. To start, let's look at the basic properties of air.

Both the density and temperature of air change with altitude above sea level. In fact, there is even a standard atmosphere model that describes average conditions and is used by aerospace engineers to predict performance of aircraft at different altitudes. For our purposes, though, we can consider sea level values as being essentially correct. The standard density of air is 1.23 kg/m^3, equivalent to a weight of approximately 0.077 lb/ft^3. We often think of air as being almost weightless, but it isn't. Indeed, a portable compressed air tank becomes noticeably heavier when filled.

The speed of sound – the speed at which pressure waves propagate through air – is 343 m/sec (1125 ft/sec) at 20 °C (68 °F).

The nominal pressure of the atmosphere at sea level is a measure of the weight of a column of air extending from the surface to the top of the atmosphere. In metric units, atmospheric pressure is 101,300 Pa (1 Pascal = 1 Newton/meter^2). This means that a square column of air 1 m on a side reaching from sea level into

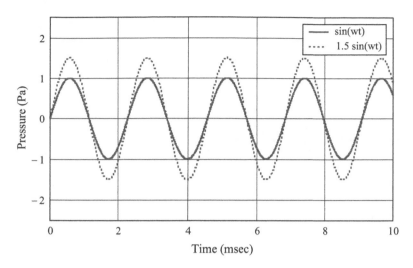

Fig. 2.2 A 10 msec Segment of Two 440 Hz Sine Waves

space weighs 101,300 N (22,773 lb). In the traditional English system, standard atmospheric pressure is 14.7 lb/in^2. This means that a square column of air 1 inch on a side reaching from sea level into space weighs 14.7 lb.

Pressure waves that we perceive as sound are superimposed over atmospheric pressure. It is perhaps surprising to find how small these pressure variations are. A pure tone (sinusoidal wave) whose peaks are ± 2 Pa is quite loud. This means that our ears are sensitive enough to easily detect deviations in sound pressure less than 0.002% of atmospheric pressure. Indeed a tone whose peaks are ±45 Pa (0.044% of atmospheric pressure) would be unbearable for most people.

Not surprisingly, there are some common measures that describe volume. The two most important are sound pressure and sound pressure level (SPL). Sound pressure is just pressure described using the RMS (root mean square) average. The idea of an RMS average is important when dealing with acoustics or electronics, so it's worth some space here.

The most common type of average is the mean. To find the mean of a list of numbers, one just adds them up and divides the sum by the number of entries in the list. For example, the mean of the list of numbers 1, 2, 3, 3, 1 is 2 since the sum of the numbers is 10 and there are five numbers in the list.

The mean is not very useful when dealing with numbers that describe a quantity oscillating about zero. Figure 2.2 shows a short segment of a sine wave (a pure tone) with a half-amplitude of 1 Pa and a frequency of 440 Hz along with the same sine wave with its amplitude increased 50%.

The second tone will be louder than the first since its volume has clearly been increased. However the mean value of the tone is zero both before and after the volume was increased. This might seem counter-intuitive. However, it just means that the curve tracing out the pressure as a function of time is symmetric about the zero pressure line (which is also the Time axis on the plot). The failure of the

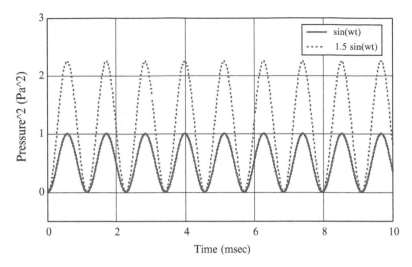

Fig. 2.3 A 10 msec Segment of Two 440 Hz Sine² Waves

mean to account for the change in volume of the tone suggests that we need a different way of describing the amplitude of sinusoidal signals; that different way is RMS.

Since the sine wave in Fig. 2.2 is symmetric about the zero pressure line, adding up the values of the points used to make the plot will always give zero. To be strictly correct, the sum will be exactly zero only if you add the values used to draw an integer number of waves, but that doesn't change the fundamental point. However, if the numbers used to draw the plot are all squared, they will all be positive and will add up to some positive number; there will be no negative numbers to cancel out the positive ones. Figure 2.3 shows the same segment of pressure data with the values squared.

From these two plots, it is easy to see that the averages must be different. In fact, the mean of the initial squared wave is 0.5 and the mean of the second one is 1.125. The only problem now is that the increase isn't proportional. If the mean of the first squared wave is 0.5 and the amplitude of the second wave (before being squared) was 50% higher, then it would be useful if the number we used to describe the volume also increased 50%. However, it went up 125%.

This problem is fixed by taking the square root of the average of the squared wave. The name Root Mean Square is almost like the instruction for how to calculate RMS. The RMS of the first tone is 0.707 and the RMS of the second, louder tone is 1.061. The second value is now 50% more than the first.

This is the reason that amplitudes of sinusoidal functions like dynamic air pressures that form music and the time dependent electrical signals that come from electric guitar pickups are expressed in RMS. For example, speakers are rated in Watts RMS and sound pressure is expressed in Pascals RMS.

Table 2.6 Environmental
Sound Pressure Levels

Sound	SPL (dB)
Permanent deafness after short exposure	150
Custom car with 1,000 watt stereo, interior	140
Pain Threshold	130
Jet takeoff at 100 m	120
OSHA max allowable, any duration	115
Front rows of rock concert	110
Outdoor rock concert	105
Large Orchestra	98
OSHA max allowable, 8 hours	90
OSHA 'action level' for 8 hour exposure	85
Car interior at highway speed	70
Conversational speech	60
Private office	50
Bedroom at night	30
Whisper	20
Threshold of hearing for a child	0

Sound pressure level, usually shortened to SPL, is related to RMS, but goes one step further. SPL is always expressed in decibels, or dB. The expression for dB is

$$dB = 20 \log \left(\frac{P_{RMS}}{P_{ref}} \right) \tag{2.2}$$

Note that the pressure is expressed in RMS. Also, the reference pressure, P_{ref}, has the same units as P_{RMS}, so dB is unitless. It is only possible to find dB if the reference value is known. By international standard, $P_{ref} = 20 \times 10^{-6}$ Pa $= 20$ μPa [2]. This value is accepted as the nominal threshold of human hearing – humans can't hear a tone with an RMS sound pressure less than 20 μPa. As one might expect, it takes more than one number to describe the lower volume limit for human hearing, but this number is roughly correct and appropriate for a reference pressure.

One nice consequence of setting $P_{ref} = 20$ μPa is that a sound of this RMS amplitude has an SPL of 0. Thus, sounds with positive SPL are loud enough to be heard (at least for a person with good hearing and in good listening conditions) and sounds with negative SPL are not. Table 2.6 shows a wide range of sound pressure levels. A sustained sound above about 100 dB is unusual in everyday life.

Since SPL, expressed in dB is logarithmic, it doesn't increase linearly as volume increases. This is actually convenient because we perceive volume in essentially the same way. The quietest tone a person could be expected to hear has an RMS value 20×10^{-6} Pa or 2×10^{-5} Pa. Conversely, it is not unusual to be exposed to noises with an RMS value of 2 Pa. This is a factor of 100,000, a very large range. In dB, this huge range is much more manageable; 20×10^{-6} Pa corresponds to 0 dB and 2 Pa corresponds to 100 dB.

2.3   String Vibrations – The Ideal String

Vibrating strings have been used to make music throughout human history. Their behavior was studied in an organized way by Pythagoras in the 6th century BCE. He was the first to observe that, if tension is held constant, then pleasing combinations of notes can be formed by changing the length of the string in specific ways. Much later, in the 17th century, Marin Mersenne discovered experimentally the relationship between fundamental frequency, length and cross-sectional area [3]

$$f \propto \frac{1}{L}\sqrt{\frac{T}{A}} \tag{2.3}$$

Later, mathematicians Jean D'Alembert and Daniel Bernoulli were able to derive and solve an equation describing string vibration starting from fundamental physical principles [4]. The solution to that equation (now called the 1-D wave equation) is

$$f = \frac{1}{2L}\sqrt{\frac{T}{\rho}} \tag{2.4}$$

Where f is the frequency in Hz, L is the length of the string, T is the tension applied to the string and ρ is the mass per unit length (sometimes called running mass). With this equation in hand, it is now easy to calculate needed information. Table 2.7 shows some useful expressions derived from Equation 2.4.

It is important to note that some assumptions were made on the way to arriving at these equations. In order to keep the math simple, it was assumed that the string has no bending stiffness at all. Rather, stiffness (resistance to bending) comes only from the tension on the string. This assumption greatly reduces the complexity of the result, but it is only an approximation. Another assumption is that tension in the string doesn't change with time as it vibrates. A third assumption is that the magnitude of the deformations is small compared to the length of the string. The result of these three assumptions is the ideal string. The mathematical expressions describing the ideal string are easy to work with and very useful for designing guitars.

Table 2.7 Expressions for Ideal Vibrating Strings

To Find	Use This Expression
Frequency	$f = \frac{1}{2L}\sqrt{\frac{T}{\rho}}$
String Length	$L = \frac{1}{2f}\sqrt{\frac{T}{\rho}}$
Tension	$T = 4f^2L^2\rho$
Mass/Unit Length	$\rho = \frac{T}{4f^2L^2}$

Since instrument designers routinely need to calculate frequencies and string tensions, it helps to show some sample calculations. Say that a steel string with a diameter of 0.254 mm (0.010 in) and a length of 647.7 mm (25.5 in) is to be tuned to 329.6 Hz. This corresponds to a typical high E string on an electric guitar.

The string is made of steel that has a density of about 7850 kg/m^3. It has a round cross-section, so the area is $\pi r^2 = \pi \times (0.000127\ m)^2 = 5.0671 \times 10^{-8}\ m^2 = 0.05067\ mm^2$. The mass per unit length is the product of material density and cross-sectional area

$$\rho = \rho_{steel} A = 7850 \frac{kg}{m^3} \times 5.0671 \times 10^{-8} m^2 = 3.978 \times 10^{-4} \frac{kg}{m} \qquad (2.5)$$

Here are some sample calculations to show how these expressions can be used:

Finding Required Tension: If the string has a length of 647.7 mm (25.5 in) and is to be tuned to high E, then f = 329.6 Hz. Then tension is

$$T = 4f^2 L^2 \rho$$

$$= 4 \times \left[329.6 \frac{1}{sec} \right]^2 \times (0.6477m)^2 \times 3.978 \times 10^{-4} \frac{kg}{m}$$

$$= 72.512 \frac{kg - m}{sec^2} = 72.512\,N = 7.39kg = 16.3lb \qquad (2.6)$$

Note that the unit of force in the metric system is the Newton, abbreviated as N (1 N = 1 kg-m/sec^2). The kilogram (kg) is a unit of mass rather than force, and reporting tension in kg is not strictly correct. However weights and forces are routinely listed in kg. Go to a grocery store in Europe and try to buy 10 N of apples. Indeed, I don't recall ever seeing Newtons used in anything but technical calculations. It is assumed that we're talking about the force corresponding to 1 kg at the surface of the Earth.

Since the f = ma and the acceleration of gravity is 9.81 m/sec^2, 1 kg at sea level corresponds to 9.81 N. At this writing, the American guitar industry still works in decimal inches. However, the metric system is used almost universally in international business. There are 4.448 Newtons in a pound, so a Newton is about the weight of a small apple. It's also about the weight of six Fig Newton cookies. There are 2.205 pounds in a kilogram at the surface of the Earth.

American readers may, at first, prefer to use traditional English units (pounds, inches and seconds) for string calculations. This creates problems since the mass unit in the English system, the slug, is defined in terms of pounds, seconds and feet. Unless you want to express the string cross-sectional area in square feet, you would have to work with a mass unit of lb-sec^2/in (sometimes informally called a slinch). This is possible, but very cumbersome – much better to work in metric units and convert to English units at the end if necessary.

Fortunately, at least one manufacturer lists tensions on their string packages and does so in both English and metric units (Fig. 2.4).

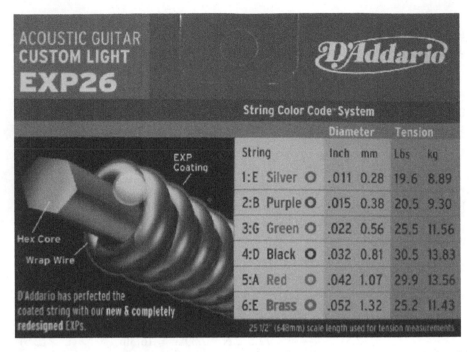

Fig. 2.4 String Packaging Showing Dimensions and Tensions (Image by the author, reproduced here courtesy of D'Addario & Co., http://www.daddario.com)

What if the string were to be installed on a guitar with a shorter scale length, say 628.65 mm (24.75 in)? Simply change the scale length and repeat the calculation

$$T = 4 \times \left[329.6 \frac{1}{\text{sec}} \right]^2 \times (0.62865\text{m})^2 \times 3.978 \times 10^{-4} \frac{\text{kg}}{\text{m}}$$
$$= 68.309\,\text{N} = 6.964\text{kg} = 15.36\text{lb} \tag{2.7}$$

Intuition says that reducing the diameter of the string should reduce the tension required to bring it to pitch. Say the string length is 647.7 mm (25.5 in) and diameter is reduced to 0.229 mm (0.009 in). The mass per unit length is reduced

$$\rho = \rho_{steel} A = \rho_{steel} \times \frac{\pi}{4} D^2$$
$$= 7850 \frac{\text{kg}}{m^3} \times \frac{\pi}{4} (0.0002286m)^2$$
$$= 3.222 \times 10^{-4} \frac{\text{kg}}{m} \tag{2.8}$$

Because the mass is reduced, intuition suggests that it will take less tension to bring the string to the correct frequency. To be sure, simply substitute the new mass value in Equation 2.6 and calculate the corresponding tension.

$$T = 4 \times \left[329.6 \frac{1}{\sec} \right]^2 \times (0.6477\text{m})^2 \times 3.222 \times 10^{-4} \frac{\text{kg}}{\text{m}}$$

$$= 58.73\text{N} = 5.98\,\text{kg} = 13.20\text{lb}$$

(2.9)

This is now enough information to calculate the fret locations on the neck. When strings are pressed against a fret, that fret becomes the end of the string, effectively shortening it. The fundamental frequency of the shortened string is calculated using Equation 2.4 just like it is for the open string. Every successive fret on the neck should raise the fundamental frequency of the string by one half step.

The expression for length of a string in Table 2.7 is used to calculate fret positions. We'll start by looking at the frequency ratio of two adjacent notes, say of an open string, f_0, and that same string at the first step, f_1.

$$r = \frac{f_1}{f_0} = \frac{\frac{1}{2L_1}\sqrt{\frac{T}{\rho}}}{\frac{1}{2L_0}\sqrt{\frac{T}{\rho}}} = \frac{L_0}{L_1}$$

(2.10)

From here, a pattern emerges

$$L_1 = \frac{L_0}{r}$$

$$L_2 = \frac{L_1}{r} = \frac{L_0}{r^2}$$

$$L_3 = \frac{L_2}{r} = \frac{L_0}{r^3}$$

$$\vdots$$

$$L_n = \frac{L_0}{r^n}$$

(2.11)

The only problem with this way of expressing fret locations is that they are measured from the nominal saddle position. For several reasons, it is much more convenient to measure fret positions from the nut. This simply requires subtracting the previous expressions from the scale length.

$$L_1 = L_0 - \frac{L_0}{r} = L_0\left(1 - \frac{1}{r}\right)$$

$$L_2 = \frac{L_1}{r} = \frac{L_0}{r^2} = L_0\left(1 - \frac{1}{r^2}\right)$$

$$L_3 = \frac{L_2}{r} = \frac{L_0}{r^3} = L_0\left(1 - \frac{1}{r^3}\right)$$

$$\vdots$$

$$L_n = \frac{L_0}{r^n} = L_0\left(1 - \frac{1}{r^n}\right)$$

(2.12)

Table 2.8 Fret Locations for Several Scale Lengths

Fret	22 in	558.8 mm	24 in	609.6 mm	24.75 in	628.7 mm	25.5 in	647.7 mm
0	0.000	0.0	0.000	0.0	0.000	0.0	0.000	0.0
1	1.235	31.4	1.347	34.2	1.389	35.3	1.431	36.4
2	2.400	61.0	2.618	66.5	2.700	68.6	2.782	70.7
3	3.500	88.9	3.818	97.0	3.938	100.0	4.057	103.1
4	4.539	115.3	4.951	125.8	5.106	129.7	5.261	133.6
5	5.519	140.2	6.020	152.9	6.208	157.7	6.397	162.5
6	6.444	163.7	7.029	178.5	7.249	184.1	7.469	189.7
7	7.317	185.8	7.982	202.7	8.231	209.1	8.481	215.4
8	8.141	206.8	8.881	225.6	9.158	232.6	9.436	239.7
9	8.919	226.5	9.730	247.1	10.034	254.9	10.338	262.6
10	9.653	245.2	10.530	267.5	10.860	275.8	11.189	284.2
11	10.346	262.8	11.286	286.7	11.639	295.6	11.992	304.6
12	11.000	279.4	12.000	304.8	12.375	314.3	12.750	323.8
13	11.617	295.1	12.673	321.9	13.070	332.0	13.466	342.0
14	12.200	309.9	13.309	338.1	13.725	348.6	14.141	359.2
15	12.750	323.9	13.909	353.3	14.344	364.3	14.779	375.4
16	13.269	337.0	14.476	367.7	14.928	379.2	15.380	390.7
17	13.759	349.5	15.010	381.3	15.479	393.2	15.948	405.1
18	14.222	361.2	15.515	394.1	16.000	406.4	16.484	418.7
19	14.658	372.3	15.991	406.2	16.491	418.9	16.990	431.6
20	15.070	382.8	16.440	417.6	16.954	430.6	17.468	443.7
21	15.459	392.7	16.865	428.4	17.392	441.8	17.919	455.1
22	15.826	402.0	17.265	438.5	17.805	452.2	18.344	465.9
23	16.173	410.8	17.643	448.1	18.195	462.1	18.746	476.1
24	16.500	419.1	18.000	457.2	18.562	471.5	19.125	485.8

Setting this up in a spreadsheet is not difficult. Table 2.8 shows the calculated fret positions for four different scale lengths.

This table suggests a way of determining the scale length of an instrument. Say you are working on a guitar that seems to be a bit smaller than normal. Simply measure the distance from the nut to the 12th fret and double it to find the scale length. This works because the 12th fret is the center point of the string.

2.4   Harmonics and Vibrational Shapes of the Ideal String

Equation 2.4 describes only the first resonant frequency of a stretched string. However a string, like any other structure, has many resonant frequencies. A convenient property of strings is that the higher resonant frequencies are multiples of the fundamental one. A more general expression for the frequencies of an ideal string is

$$f_n = \frac{n}{2L} \sqrt{\frac{T}{\rho}}$$

$$(2.13)$$

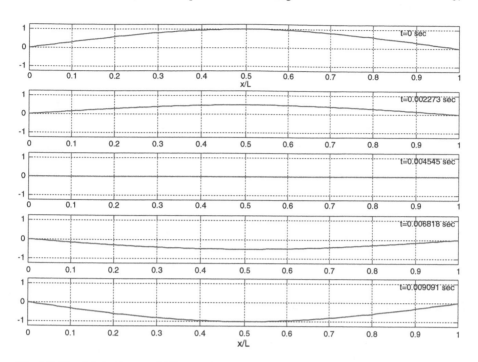

Fig. 2.5 First Mode Shape at Different Times

Where n is an integer, n = 1, 2, 3, ... This means that, if the string was tuned to a fundamental frequency of 110 Hz, then the second resonant frequency, the first harmonic, would be 220 Hz. The third resonant frequency, the second harmonic, would be 330 Hz and so on.

Each of the resonant frequencies of a string has a corresponding vibrational shape, called a mode shape [4]. Take our string tuned to vibrate at 110 Hz. If a small electromagnet was placed near the string and driven with a 110 Hz sine wave (technically, the signal would need to have an offset so that the minimum voltage was zero, but that's not important for this discussion), it would vibrate with the first mode shape. If you were to illuminate the vibrating string with a strobe light, freezing the motion of the string at different points in its motions, the result would look something like Fig. 2.5.

Figure 2.5 shows the first mode of the string, but subsequent mode shapes are closely related to the first one. The first mode shape of the ideal string is just the first half of a sine wave. The second mode shape is just a whole sine wave. The third mode shape is 1.5 sine waves and so on. Figure 2.6 shows the first four mode shapes. These are what the string would look like if high speed flash pictures were taken at the time of maximum deflection. To excite the second mode, just drive the string at twice the fundamental frequency. To excite the third mode, drive it at three times the fundamental and so on.

The previous two figures show how a string responds when driven at a resonant frequency. While this is certainly possible, it is not something that generally happens outside a lab environment. Rather, strings are plucked and allowed to vibrate. For reasons that involve some heavy math, plucking a string excites many different resonant frequencies at once.

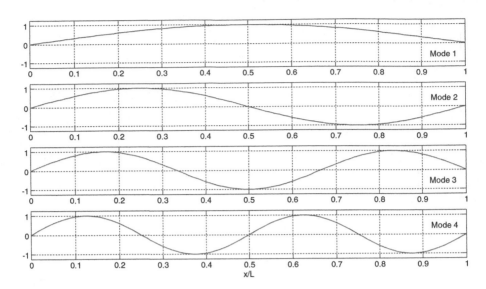

Fig. 2.6 First Four Mode Shapes of an Ideal String

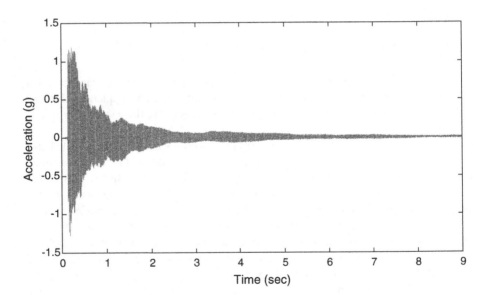

Fig. 2.7 D String Response in Time Domain

Figure 2.7 shows the response measured from an acoustic guitar when the D string is plucked. This data was recorded using an acceleration sensor (an accelerometer) fixed to the top, and acceleration is plotted as a function of time. Physics tells us that there must be many frequencies present in this recorded signal, but it's impossible to tell from looking at this plot.

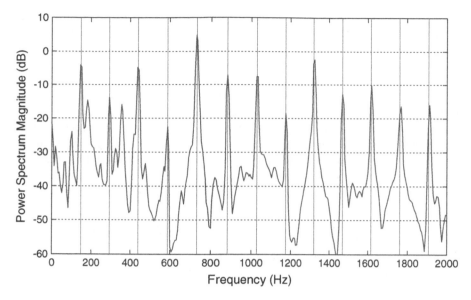

Fig. 2.8 D String Response in Frequency Domain

The helpful alternative is to look at it in frequency domain. We live in time domain and all test data is recorded in time domain, like that in Fig. 2.7. However, it is not always helpful to look at the data in time domain. A mathematical tool called the Fourier transform allows us to take data that has been recorded in time domain and transform it into frequency domain. This way, it can be plotted against frequency rather than time. The computer implementation of the Fourier transform is call the Fast Fourier Transform and is usually abbreviated as FFT. FFT functions are built into most frequency processing software and a basic version is even built into Excel.

It's important to understand that transforming data into frequency domain doesn't change the information present. This is an abstract concept, but we transform numbers all the time without being too fussed. International travelers might change their dollars into Euros or yen; while the money might look different after the exchange, it is still, in a most fundamental sense, the same amount of money. Let's assume, for the sake of simplicity, that the transaction is free, so nothing is lost to service fees. Just as a traveler coming back to the United States from Europe might change Euros back to dollars without losing money, an analyst can transform frequency domain back to the time domain without losing information.

Figure 2.8 shows the same string data in frequency domain. Now, it is very easy to see that the fundamental frequency of 146.8 Hz is present along with a number of higher frequencies – harmonics. The vertical lines are at multiples of the fundamental frequency; the first vertical line is the fundamental (n = 1), the second vertical line is the first harmonic (n = 2) and so on.

For clarity, only the frequencies up to 2000 Hz are shown here. The data set does, however, contain harmonics through 24,000 Hz (24 kHz). The human hearing range is generally assumed to be from 20 Hz to 20 kHz, so this string is making sounds of such high frequency that no human could hear them.

There are some other interesting features in this plot. One is that the heights of the peaks are all different. This indicates that not all resonant frequencies of the string are present in the same amount. For example, the resonant peak near 600 Hz (actually 587.3 Hz) is very low compared to most of the others. This means that the motion of the body where the sensor was placed did not have a strong component at this frequency. It would not be surprising if the resulting sound had little content at this frequency. Conversely, the next peak, at 734.2 Hz, is a very strong one and it would not be surprising if the resulting sound had a strong component at this frequency.

One other feature in this plot is that there are peaks that can't be explained by the string vibration alone. These are from the vibration of the rest of the guitar structure. The body of the guitar has its own resonant frequencies that are determined by both the structure of the instrument and the air enclosed in the body.

2.5 Real Strings

The previous section describes ideal strings. The ideal string is easy to describe mathematically, but it is only an approximation; we don't get to make guitars with ideal strings. Real strings have properties that modify their response from the ideal and these differences must be accounted for in the design of guitars [5].

There are three major factors that could be taken into account in order to refine the expression for ideal strings: variable tension, large displacements and non-zero bending stiffness. For guitar strings, the displacement really is small compared with length. As a result, the tension really is approximately constant [6]. Rather, the assumption of zero bending stiffness appears to be a major source of disagreement between ideal predictions and the behavior of real strings.

In order to account for bending stiffness, there needs to be a way of accounting for the stiffness of the material from which the string is made. The number describing material stiffness is the elastic modulus of the material, E. It is also sometimes called Young's modulus, after Thomas Young (1773-1829), an English polymath who, among other things, also developed his own method of temperament. In older technical references, the elastic modulus is sometimes designated as Y.

Elastic modulus has units of pressure (force/length^2) and the numbers are large. For common forms of steel, $E = 205 \times 10^9$ Pa $= 29.7 \times 10^6$ lb/in^2. The elastic modulus of nylon depends on its formulation, but we can assume an approximate value of $E = 2 \times 10^9$ Pa $= 290 \times 10^3$ lb/in^2.

The equation that describes the behavior of a vibrating string and includes bending stiffness is very difficult to solve, so the published solutions are approximate. Fortunately for us, the approximate solutions are fairly accurate, certainly enough so to be useful here.

There are two particularly simple expressions in the literature for the frequencies of strings with bending stiffness [7,8]. The simpler of the two is

$$f_n = \frac{n}{2L}\sqrt{\frac{T}{\rho}}\sqrt{1 + \frac{n^2\pi^3 E r^4}{4L^2 T}} \tag{2.14}$$

Where n indicates the nth resonant frequency. Also, ρ is the mass per unit length, E is the elastic modulus of the string material, T is tension, L is length and r is the radius of the string. The expression contained in the second square root sign describes inharmonicity since it simply modifies the expression for the ideal frequencies. It is helpful to clearly define the expression for inharmonicity

$$I_n = \sqrt{1 + \frac{n^2\pi^3 E r^4}{4L^2 T}} \tag{2.15}$$

If $I_n = 1$, then there is no inharmonicity. If the goal is to reduce inharmonicity, this equation suggests some options. Since elastic modulus and radius are in the numerator, reducing them will reduce inharmonicity. Since string length and tension are in the denominator, increasing them will decrease inharmonicity.

Another expression for string resonant frequencies that includes the effect of bending stiffness is

$$f_n = \frac{n}{2L}\sqrt{\frac{T}{\rho}}\left[1 + \frac{r^2}{L}\sqrt{\frac{\pi E}{T}} + \left(4 + \frac{n^2\pi^2}{2}\right)\frac{\pi E r^4}{4TL^2}\right] \tag{2.16}$$

As before, the expression in braces simply modifies the ideal expression for string resonant frequencies, so it expresses inharmonicity. If the elastic modulus is zero, then the string has no bending stiffness and the result is the ideal expression in Equation 2.13. The definition of inharmonicity from this expression is

$$I_n = 1 + \frac{r^2}{L}\sqrt{\frac{\pi E}{T}} + \left(4 + \frac{n^2\pi^2}{2}\right)\frac{\pi E r^4}{4TL^2} \tag{2.17}$$

As before, decreasing elastic modulus and radius, and increasing length and tension all decrease inharmonicity.

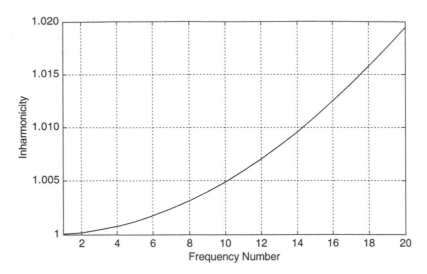

Fig. 2.9 Predicted String Inharmonicity

In order to make these expressions more real, it helps to have a concrete example. Consider a steel G string with a diameter of 0.4064 mm (0.0160 in) and a length of 647.7 mm (25.5 in). At concert pitch, the G string is tuned to 196 Hz. The ideal string expression predicts the tension to be 65.64 N (14.757 lb).

Both Equations 2.10 and 2.12 predict slightly higher fundamental resonant frequencies than the ideal string equation. Thus, if the string is to be at concert pitch, the tension must be slightly reduced. Equation 2.14 predicts that the actual tension required to bring the string to the correct tension is very slightly less than the ideal prediction, 65.508 N (17.727 lb). Equation 2.16 predicts the effect of string bending stiffness is more pronounced and that, accordingly, the tension required to bring the string to the correct pitch is lowered more significantly at 64.678 N (14.540 lb).

Figure 2.9 shows a plot of the inharmonicity equation for the steel G string. Note that, if the corrected tensions are applied, both equations 2.14 and 2.16 give the same prediction and either one could be used to produce this plot.

The closer inharmonicity is to 1, the more the real string behaves like an ideal one. This is desirable because the higher frequencies of an ideal string are integer multiples of the fundamental and that means they are separated by octaves. When there is inharmonicity, the higher frequencies are not quite integer multiples of the fundamental frequency and may sound slightly dissonant. However, sometimes small changes can help reduce inharmonicity; an example or two will help show how changes in string parameters affect inharmonicity.

To start, let's consider the example of the 0.4064 mm (0.0160 in) diameter G string tuned to 196 Hz. The density of steel is 7850 kg/m^3, so the mass per unit length of the string is $\rho = 1.016 \times 10^{-3}$ kg/m. First, let's consider a string where L = 647.7 mm (25.5 in).

Table 2.9   Calculated Effects of Geometry Changes on Inharmonicity of a G String

Frequency Number, n	Nominal String	Increased Length	Decreased Diameter	Decreased Diameter and Increased Length
	T = 65.64 N	T = 76.672 N	T = 50.258 N	T = 58.702 N
1	1.00005	1.00004	1.00004	1.00003
2	1.00020	1.00014	1.00015	1.00011
3	1.00044	1.00032	1.00034	1.00025
4	1.00079	1.00058	1.00060	1.00044
5	1.00123	1.00090	1.00094	1.00069
6	1.00177	1.00129	1.00135	1.00099
7	1.00241	1.00176	1.00184	1.00135
8	1.00314	1.00230	1.00241	1.00177
9	1.00398	1.00291	1.00305	1.00223
10	1.00491	1.00359	1.00376	1.00276
15	1.01101	1.00808	1.00844	1.00619
20	1.01949	1.01432	1.01495	1.01098

Let's say that the tension predicted by the ideal string equation is 65.64 N (14.757 lb). The second column of Table 2.9 shows the predicted inharmonicity of the unmodified string using Equation 2.15. The third column shows the effect of increasing the length of the string to 700 mm (27.57 in). This scale length would be extremely long for a guitar, though it would be typical for a baritone guitar. Note that, to keep the fundamental frequency of 196 Hz, the ideal string equation predicts that tension must be increased to 76.672 N (17.236 lb).

The fourth column shows the predicted effect of decreasing the diameter to 0.3556 mm (0.0140 in). Note that the smaller diameter reduces the predicted tension to 50.258 N (11.298 lb). The decreased diameter and decreased tension have opposite effects on inharmonicity, so it's not obvious what the result would be. It turns out that decreasing string diameter reduces inharmonicity, though fractionally less than the previous increase in string length.

Finally, the fifth column shows the effect of both increasing length and decreasing string diameter. Combining these two changes significantly reduces inharmonicity.

It's generally easier to see trends graphically, so the data in Table 2.9 is also plotted in Fig. 2.10. The significant decrease in inharmonicity is clearly visible.

2.6   Measurements from Real Strings

The last step in examining the behavior of strings is to compare predictions for real strings to actual measurements. In an actual test, there are several ways that a string might behave differently than mathematical predictions. One, of course, is inharmonicity. Another, however, is the end conditions of the string. Both the

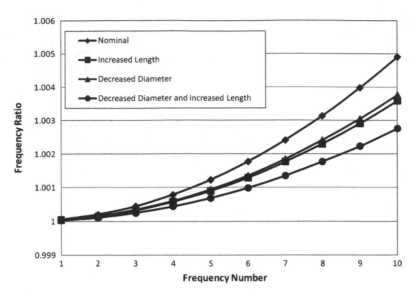

Fig. 2.10   Effect of Geometry Changes on String Inharmonicity

ideal string equation (Equation 2.9) and the two real string equations (Equations 2.10 and 2.12) assume that the ends of the string are fixed to a rigid, massive structure. It has been shown analytically that fixing the ends of the strings to a flexible structure, like a guitar, does affect their behavior. This slight additional flexibility at the ends of the strings might cause a measurable departure from predictions based on ideally fixed ends [9].

In order to make a clear comparison between predictions and test results, a heavy steel fixture was made to hold a single string so its motion could be measured with a laser vibrometer. Weight is on the order of 20 kg (44 lb) and the string was supported by heavy steel saddles welded to the supporting bar. Figure 2.11 shows one end of the fixture with the laser dot illuminating the string.

The laser vibrometer measures the velocity of the string without touching it or affecting its motion with a magnetic field. The string was tuned to G (196 Hz) and plucked several times while the vibrometer recorded the response. The resulting inharmonicity data is shown in Fig. 2.12 along with a curve fit. The inharmonicity increases with the square of the frequency number to a very close approximation. Also, the test data agrees very well with the analytical data shown in Fig. 2.9.

2.7   Correcting for Inharmonicity

It's clear that resonant frequencies of real strings are higher (sharper) than for ideal strings. Shortening the string by pressing it against a fret creates inharmonicity for two reasons [10]. The first is that decreasing length increases inharmonicity due to

Fig. 2.11 String Test Fixture

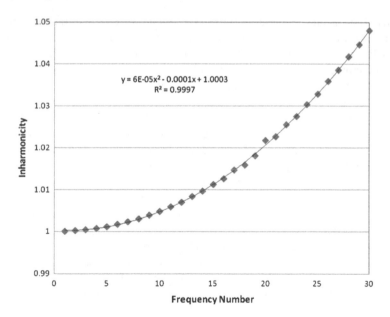

Fig. 2.12 Inharmonicity Measured from Test Fixture

the increased effect of bending stiffness. The second is that the string is stretched slightly by being pushed against the fretboard, a distance on the order of 3 mm (about 1/8 inch) at the 12th fret, so the tension increases. Figure 2.13 shows the basic geometry of a string being stretched as it is pushed down against the fretboard.

Fig. 2.13 A String Being
Stretched as it is Pressed
against the Fretboard

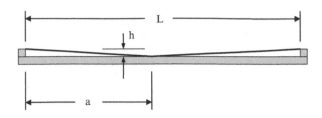

The length of the stretched string is

$$L + \Delta L = \sqrt{a^2 + h^2} + \sqrt{(L - a)^2 + h^2} \qquad (2.18)$$

The change in tension comes from the definitions of stress and strain

$$\Delta T = EA\frac{\Delta L}{L} = \frac{EA}{L}\left(\sqrt{a^2 + h^2} + \sqrt{(L - a)^2 + h^2} - L\right) \qquad (2.19)$$

Where E is the elastic modulus of the string material, a number that describes its resistance to stretching. Unless some sort of correction is built into the instrument, notes will sound increasingly sharp as the player moves up the neck. There are a number of approaches that would work, at least in theory, but the universal solution has been to move the saddle(s) slightly farther from the nut than the actual scale length requires.

This has the effect of making the string slightly longer at each fret than the ideal string equation requires, thus slightly reducing its pitch. The process of setting the saddle position is called intonation or intonating. Typically, the string is tuned to the correct open pitch and then played at the 12th fret. If the note is sharp, the saddle is moved a little farther from the nut. The open string tuning is corrected and the process is repeated. When the string is at the correct pitch both open and at the 12th fret, it has been intonated. The process is repeated for the other five strings.

Of course, when using this method of pitch correction, the string is only at the correct pitch open and at the 12th fret. However, it is close enough at the other frets that the error is generally not perceptible. Figure 2.14 shows the bridge on the author's Squier Stratocaster with saddles adjusted using the open-12th fret method.

Acoustic bridges need to be light and seldom have individually adjustable saddles. Rather, they almost always have straight, one-piece saddles. On steel string acoustic guitars, the saddle is generally offset so that the high E string has very little intonation and the low E has the maximum. Sometimes, there are notches in the saddle to make corrections for individual strings. Most typical is a notch for the B string. Figure 2.15 shows the bridge on a Fender CD-60 acoustic guitar with an offset saddle and a notch for the B string.

Fig. 2.14 A Stratocaster Bridge with Adjustable Saddles

Fig. 2.15 An Acoustic Bridge with Offset Saddle

The required intonation depends strongly on the elastic modulus of the string material. The elastic modulus of nylon is much less than that of steel, so classical guitars need much less intonation. It is typical for the saddles to be perpendicular to the strings and simply offset by a small amount. A perpendicular saddle is shown on the bridge of a classical guitar in Fig. 2.16. There is no standard value, but offsets on the order of 2.5 mm (0.1 in) are typical.

Fig. 2.16 A Classical Guitar
Bridge with Saddle
Perpendicular to the Strings
(Image by the author,
reproduced here courtesy
of Prof. Eugene Coyle)

2.8   Physics of Guitars

Vibrating strings alone make very little sound because they have very little surface
area. A vibrating object can only radiate sound if its motion causes pressure waves
in the surrounding air. There are essentially two ways this happens with guitars.
Acoustic guitars use vibration of a thin soundboard to make the pressure waves we
perceive as sound. Electric guitars use a sensor called a pickup that detects motion
of the strings and converts that motion to a dynamic electrical signal. That signal is
amplified and used to drive a speaker that, in turn, makes pressure waves. These two
ways of turning string vibrations into sound are different enough that it makes sense
to address them separately.

2.8.1   Acoustic Guitars

The familiar shape of the acoustic guitar is so much a part of popular culture that it
is easy to forget it is a mechanism for transforming string motion into sound. In
more precise terms, it converts the kinetic energy of the strings into kinetic energy
in the structure of the guitar. The guitar has a large surface area so it is able to
transfer the kinetic energy in the body into the air in the form of pressure waves.

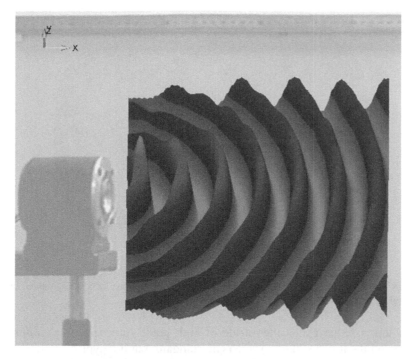

Fig. 2.17 A 10 kHz Sound Wave Produced By A Speaker (Image courtesy of Prof. Tom Huber, Department of Physics, Gustavus Adolphus College)

Perhaps the most familiar example of a moving surface creating a sound wave is a speaker cone. Figure 2.17 shows a remarkable image of a sound wave measured directly by a device called a scanning laser vibrometer[11]. The small speaker at the left side of the picture is making a 10 kHz pure tone so the wavelength is approximately 33.8 mm (1.33 in).

Any moving surface can radiate sound, including the surface of a guitar body. To be precise, the pressure created in the surrounding air is proportional to the velocity of the moving surface.

Another way to produce sound is drive a volume of air at a resonant frequency so that the air itself vibrates by compressing and expanding. Anyone who has ever made a whistle by blowing over the open top of a bottle has demonstrated this effect. The turbulent air flowing past the top of the bottle excites the resonant frequencies in the confined air, though the first resonant frequency is usually the dominant one. The frequency of the radiated sound can be varied by adding some liquid to the bottle to change its volume; reducing the volume of air inside the bottle increases the resulting frequency [12]. If you haven't tried this, you should.

An acoustic guitar makes sound both by radiating it from the soundboard, in a manner similar to a speaker, and by radiation from the soundhole, like the pop bottle.

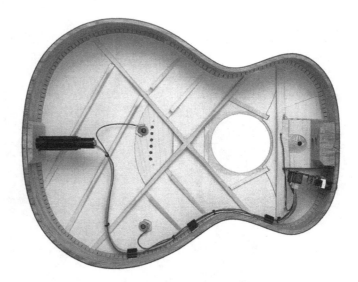

Fig. 2.18 Interior View of a Taylor Acoustic Guitar (Image courtesy of Taylor Guitars, http://www.taylorguitars.com)

The top is a thin plate, often made of high grade spruce, with a reinforcing structure on the inside surface. There is no standard, but it is typical for tops to be around 2.5 mm (0.098 in) thick. Figure 2.18 shows a cutaway of a Taylor acoustic guitar with interior bracing and electronics. The back and sides of this instrument are made from thin pieces of mahogany. Sides are often a little thinner than tops to make bending easier, and backs are often around the same thickness at the tops. The result is a structure strong enough to withstand the combined tension of the strings while still being flexible enough to vibrate in response to the vibrating strings.

A prominent feature of most acoustic guitars is the soundhole. One sometimes hears an interested person unfamiliar with the mechanics of acoustic guitars wonder if the purpose of the soundhole is to 'let the sound out'. Its actual purpose is more subtle than that since it both acts as a radiator and as a tuning port for the air in the body. The air inside the body of the guitar has both mass and stiffness and this affects the dynamic response of the instrument.

We are generally familiar with the idea that air is compressible and that a large volume of air can be made to fit into a small container if enough force is applied. For example, we've all filled a car or bicycle tire with air from a pump or compressor. The fact that it takes force to compress air means that it has stiffness. However, we are often a little less clear about how much mass air has.

At standard temperature and pressure (think pleasant day at sea level) the density of air is about 1.23 kg/m^3. In the English system, the unit for density is slug/ft^3. This is just too cumbersome, so let's settle for the weight density of 0.0765 lb/ft^3. Whatever the units, the bottom line is that air weighs enough to matter. For example, the air enclosed by a large acoustic guitar has a mass in the neighborhood of 21 gm (about 3/4 ounce).

Fig. 2.19 Schematic
of Energy Flow in a Guitar
(After Fletcher and
Rossing [9])

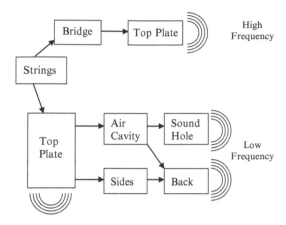

2.8.1.1 Body Vibrations

All structures have multiple resonant frequencies – frequencies they like to amplify.
For strings, those frequencies are roughly integer multiples of the lowest frequency,
usually called the fundamental. The bodies of acoustic guitars generally have a
number of resonant frequencies within the range of notes they can play. Before
focusing on the specifics of how the structures of acoustic guitars vibrate, it is
important to understand the role structural vibrations play in the sound producing
mechanisms of acoustic guitars.

Vibrating strings transmit energy to the flexible structure of the guitar. As the
structure vibrates, it radiates sound outward and excites the air volume within the
body. Fletcher and Rossing [9] presented a diagram (Fig. 2.19) that shows how
energy propagates through the guitar structure and distinguished between low
frequency and high frequency.

In mechanical terms, the strings drive the structure which responds mostly at the
string resonant frequencies, a process called forced response. This is why an
acoustic guitar radiates the resonant frequencies of the strings and can make
music. However, the structure of the guitar has its own resonant frequencies and
they color the resulting sound.

There is a wide range of acoustic guitar designs and they can be expected to have
their own dynamic characteristics. However, most guitars have roughly the same first
three mode shapes as shown in Fig. 2.20. The first mode (a) is formed by the motion of
the top and back in opposite directions so that there is a net change in volume. The
corresponding frequency is typically in the range of 95 Hz – 105 Hz. The second mode
(b) is similar except that the top and back move in the same direction and there is little
change in body volume. The frequency for this mode is typically in the neighborhood
of 200 Hz. The shape of the third mode (c) varies between instruments, but it is
typically the first one with an internal node line – a line of points that don't move.

Finding resonant frequencies is definitely easier than measuring mode shapes.
It is a straightforward process to use an instrumented hammer and an accelerometer

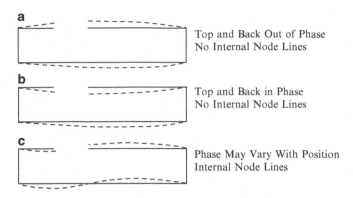

a   Top and Back Out of Phase
         No Internal Node Lines

b   Top and Back in Phase
         No Internal Node Lines

c   Phase May Vary With Position
         Internal Node Lines

Fig. 2.20 Schematic of the First Three Mode Shapes of an Acoustic Guitar Body

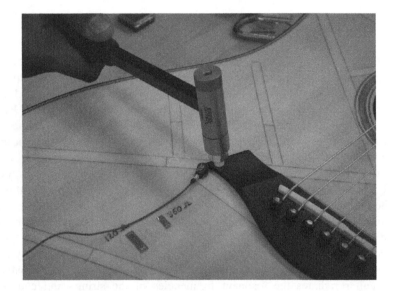

Fig. 2.21 Tap Testing a Guitar

to measure the frequency response function of a guitar top and to identify the resonant frequencies from it. Note that the accelerometer was placed right next to the point where the hammer impacted the top, so the result is called a driving point Frequency Response Function (FRF). Figure 2.21 shows the hammer and accelerometer on the top of a Taylor 710 guitar. Note also that the two strain gauges, labeled 120 Ω and 350 Ω are left over from a previous test. This instrument is unusual in that it had the bracing pattern drawn on the top before the clear finish was applied.

Figure 2.22 shows the frequency response function (FRF) measured from the guitar in Fig. 2.21. The peaks below 100 Hz are likely due to bending of the complete instrument and from the way the instrument is supported. The first body mode is at 98.1 Hz, the second is at 181 Hz and the third is at 324 Hz.

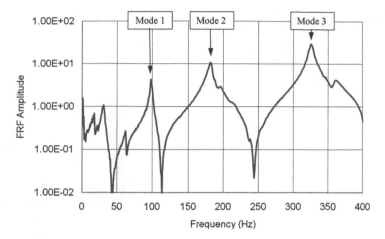

Fig. 2.22   Driving Point FRF of a Taylor 710

Fig. 2.23   Chladni Pattern of a Round Plate (Wikimedia Commons, image is in the public domain)

There are a number of ways to measure the mode shapes of a vibrating plate. One often used by guitar makers is Chladni patterns, named after Ernst Chladni, a physicist and musician whose works include research on vibrating plates [13]. A node line on a vibrating plate is a collection of points where there is no motion. If a plate is sprinkled with small particles like sand or salt and made to vibrate, the particles will collect at the node lines. Figure 2.23 shows a Chaldni pattern on a round disk that has been fixed to a small electromagnetic shaker. The driving frequency is 5670 Hz and the sand shows the node lines of a nice, symmetric mode shape. When working with guitars, the excitation usually comes from a speaker driven using a function generator and amplifier.

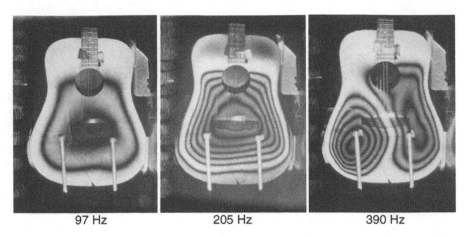

97 Hz	205 Hz	390 Hz

Fig. 2.24 Fringe Patterns Showing the First Three Mode Shapes of an Acoustic Guitar (Images courtesy of Karl Stetson, http://www.holofringe.com)

A more sophisticated means of imaging mode shapes is holography. While single exposure holography makes a 3-D image of an object, methods that combine two holographic images produce lines, called fringes, which join points of constant displacement. In this sense, they are analogous to the lines on topographic maps. Figure 2.24 shows several of a classic set of images by Karl Stetson [14] in 1972 of a dreadnought guitar. Note that the two white posts are tubes leading from speakers used to acoustically excite the guitar top.

The 97 Hz and 205 Hz shapes do not have an internal node line while the 390 Hz shape has a node line that roughly follows the bracing pattern. This progression of mode shapes matches the diagram in Fig. 2.20.

2.8.1.2 Coupling between Body and Air

The conceptual diagram in Fig. 2.19 clearly shows that the air inside the guitar body both radiates directly from the soundhole and affects the motion of the flexible top and back. This acoustic coupling is central to the way an acoustic guitar produces sound, but is sometimes difficult to understand beyond some intuitive level. Fortunately, there are some simple math models that account for this interaction between the structure and the air.

Probably the most widely cited one uses just two degrees of freedom (DOF): one for the motion of the flexible top and another for air moving in and out of the soundhole [15]. The sides and back are assumed to be rigid, though an extension of this model includes a flexible back.

The two DOF model couples a flexible top surface with what is otherwise a Helmholtz resonator as shown in Fig. 2.25. The walls are assumed to be rigid everywhere except for the flexible section. The flexible section of the top is

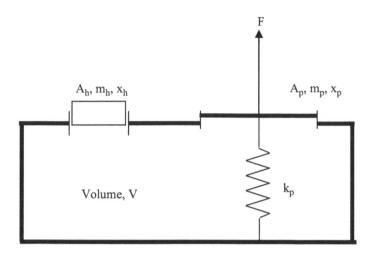

Fig. 2.25 Two Degree of Freedom Model

modeled by a rigid flat plate supported by a linear spring (thus, it cannot have any internal node lines). The two sources of stiffness are this spring and the compressible air inside the body. The two sources of mass are the movable portion of the top and the mass of the air column moving in the sound hole.

Clearly, the soundhole doesn't really have a tube in it that would create an air column, but this is one of the assumptions of a Helmholtz radiator on which this model is based. Remember that this model is approximate and simple; insisting on more accuracy brings with it much mathematical complication.

It's easy to imagine that if the air column moved down, the air in the body would be compressed and there would be an additional force against the movable portion of the soundboard. Conversely, if the soundboard moved down, the increased pressure in the body would force the air column up.

The model shown in Fig. 2.25 is defined mathematically in terms of physical parameters. In practice, these parameters are generally inferred from measurements of an existing instrument. They are:

μ	Proportionality constant between changes in volume and changes in pressure, $\mu = c^2 \rho / V$
ρ	Density of air
A_h	Area of sound hole
c	Speed of sound
m_h	Mass of the air column moving through sound hole
x_h	Position of air column moving through sound hole
V	Volume of enclosed air
k_p	Stiffness of top plate
F	Force applied to top plate
A_p	Effective area of top plate
m_p	Effective mass of top plate
x_p	Position of top plate

The goal of this model is to demonstrate the connection between the structure and the air and to calculate resonant frequencies based on representative physical parameters. The details are beyond the level of this book. However, the result is a mathematical entity called a characteristic equation.

$$m_p m_h \lambda^2 - \left(k_p m_h + \mu A_p^2 m_h + m_p \mu A_h^2\right) \lambda + \left(k_p \mu A_h^2 + \mu^2 A_p^2 A_p^2 - \mu^2 A_h^4\right) = 0$$

(2.20)

It is simply a parabola and can be written out as

$$a_2 \lambda^2 + a_1 \lambda + a_0 = 0 \tag{2.21}$$

Where

$$a_2 = m_p m_h$$
$$a_1 = -\left(k_p m_h + \mu A_p^2 m_h + m_p \mu A_h^2\right) \tag{2.22}$$
$$a_0 = k_p \mu A_h^2 + \mu^2 A_p^2 A_p^2 - \mu^2 A_h^4$$

The Greek letter lambda, λ, is related to the resonant frequency

$$\lambda = \omega^2 = (2\pi f)^2 \tag{2.23}$$

To find the two resonant frequencies of the model in Fig. 2.25, find a_0, a_1 and a_2 and solve Equation 2.21 to find two different values of λ. Then use Equation 2.23 to find the two corresponding frequencies. If the physical parameters are chosen correctly, the calculated frequencies should be close to the measured ones for a specific instrument.

Note that, if the flexible top plate is removed, then $x_p = 0$ and the result is a Helmholtz resonator. The resonant frequency is then $\omega_h = (\mu A_h^2/m_h)^{1/2}$. This should correspond to the frequency of the lowest point, the anti-resonance, between the first two body resonances. In Fig. 2.22, that point lies between the peaks at 98.1 Hz and 181 Hz and is 114 Hz.

2.8.1.3   Two DOF Example

A 2-DOF model was tuned to match the response of a Taylor steel string grand auditorium style guitar as shown in Fig. 2.26.

There are a several different methods of tuning the math model [15,16], but any tuning process requires test data. Tests showed that the first two resonant frequencies of the instrument were 99.4 Hz and 185 Hz and the Helmholz frequency was 115 Hz. By comparing test data with calculated frequencies, the model

Fig. 2.26 Taylor Grand Auditorium Acoustic Guitar (Image courtesy Taylor Guitars, http://www. taylorguitars.com)

Table 2.10 Parameters for 2-DOF Model

Parameter	Description	Value
A_h	Sound hole area	81.89 cm^2
A_p	Top plate area	630.0 cm^2
c	Speed of sound in air	338 m/s
k_p	Top stiffness	191,400 N/m
m_h	Mass of air column in sound hole	1.247 gm
m_p	Top mass	190.1 gm
V	Volume of body	14.66 L
ρ	Density of air	1.23 kg/m^3

parameters were determined to be those shown in Table 2.10. The calculated resonance frequencies using this model are 99.31 Hz and 183.8 Hz. The calculated Helmholtz frequency is 114.3 Hz.

Finding the unknown constants in the characteristic equation is a process from a field of study called parameter identification. There are eight parameters in the table above, but the speed of sound and the density of air can generally be assumed from standard tables. The speed of sound at sea level is 338 m/sec (1109 ft/sec) and the density of air at sea level is 1.23 kg/m^3 (0.00239 slug/ft^3).

The remaining six parameters can be determined using measured frequencies. However, there must be at least as many measured frequencies as there are unknown parameters and it helps to have more than six. A convenient method for generating this additional data is to measure frequencies with added mass and with the soundhole covered. A typical test might include the following conditions:

	Additional Mass	Soundhole Covered
1	No	No
2	Yes	No
3	No	Yes
4	Yes	Yes

Fig. 2.27 Mass Fixed to the
Soundboard

Since there are two measured modes per test and four test conditions, there will
be eight data points to use in identifying the unknown parameters. If the anti-
resonance between the first and second resonant frequencies can be identified, it can
be the ninth data point. A least squares error process is then used to find the model
parameters.

Figure 2.27 shows a socket of known mass (51.3 gm) attached near the center of
the moving portion of the soundboard. The mass of the socket is simply added to m_p
in order to calculate resonant frequencies for this configuration.

Figure 2.28 shows the soundhole covered with a light, stiff piece of corrugated
cardboard. To account for this configuration, $A_h = 0$.

Note that the strings are heavily damped with pieces of soft foam. This is to
ensure that the measured response is that of the guitar body rather than of the strings
and body together. Without some sort of damping material, the response is
completely dominated by the contribution of the strings.

2.8.2 Electric Guitars

The primary difference between electric and acoustic guitars is how they produce
sound. The behavior of the strings, the layout of frets and the need for intonation are
essentially the same. However, solid body electric guitars dispense with the hollow,
flexible body structure and replace it with pickups that sense string motion.

Fig. 2.28 An Acoustic Guitar with the Soundhole Covered

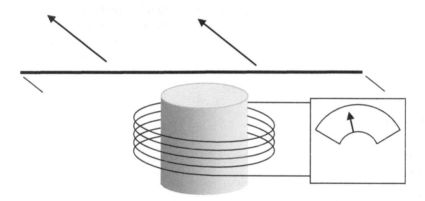

Fig. 2.29 Electromagnetic Induction

They produce a time-varying voltage that is proportional to the velocity of the strings and it is this signal that is amplified to produce sound.

The details of how pickups work are presented in Chapter 4. For now, it is enough to know that an electromagnetic pickup works by sensing movement of a magnetic field caused by motion of a string as shown in Fig. 2.29. Pickups are made using one or more magnets and a coil of wire. String motion makes the magnetic field move and this causes electrons to flow through a wire coil using a phenomenon called induction.

Inductive sensors that work just like electric guitar pickups are used in many different applications, but one of the most common is for wheel speed sensors.

Fig. 2.30 A Motorcycle Wheel Speed Sensor (Wikimedia Commons, image is in the public domain)

Anti-lock brake systems (ABS) sometimes use inductive sensors to measure the rate at which teeth on a gear pass as a means of determining wheel speed. There are a known number of teeth on the gear, so knowing the rate at which they pass the sensor means knowing the rotation speed of the wheel. A motorcycle wheel speed sensor is shown in Fig. 2.30.

References

1. ISO (1975) 16:1975 Acoustics – Standard Tuning Frequency (Standard Musical Pitch). International Standards Organization
2. ISO (2007) ISO 80000-8:2007 Quantities and Units – Part 8: Acoustics. International Standards Organization
3. Stillwell J (2010) Mathematics and its History, 3rd ed. Springer
4. Inman DE (2007) Engineering Vibration, 3rd ed. Prentice Hall
5. Young RW (1952) Inharmonicity of Plain Wire Piano Strings. Journal of the Acoustical Society of America 24:267–273
6. Keller JB (1959) Large Amplitude Motion of a String. American Journal of Physics 27:584
7. Young RW (1954) Inharmonicity of Piano Strings. Acustica 4:259–262
8. Shankland RS and Coltman JW (1939) The Departure of the Overtones of a Vibrating Wire from a True Harmonic Series. Journal of the Acoustical Society of America 10:161–166
9. Fletcher NH and Rossing TD (1998) The Physics of Musical Instruments 2nd ed. Springer
10. Varieschi GU and Gower CM (2010) Intonation and Compensation of Fretted String Instruments. Amercan Journal of Physics 78:47–55
11. Oliver DE (1995) Scanning Laser Vibrometer for Dynamic Deflection Shape Characterization of Aerospace Structures. Proceedings of the SPIE 2472:12–22
12. French RM (2005) A Pop Bottle as a Helmholz Resonator. Experimental Techniques 29:67–8
13. Ullmann D (2007) Life and Work of E.F.F. Chladni. The European ysical Journal

14. Stetson KA (1981) On Modal Coupling of String Instrument Bodies. Journal of Guitar Acoustics 3:23–31
15. Christensen O and Vistisen BB (1980) Simple Model for Low-Frequency Guitar Function. Journal of the Acoustical Society of America, 63:758–766
16. French RM (2009) Engineering the Guitar: Theory and Practice, Springer

Chapter 3
Structure of the Guitar

The design of guitars – either acoustic or electric – is strongly conditioned by the structural requirements. Any guitar must be strong enough to withstand the tension of the strings, light enough to be comfortable and shaped in a way that makes it easy to hold and to play. Acoustic guitars have the additional requirement that the soundboard and back must be flexible enough to move in response to the vibrating strings.

A guitar must also have an acceptably long service life. A bare wood guitar is not durable. It is easily damaged, can get dirt or oil worked into the grain and can absorb moisture through either direct contact or from humidity in the air. To be practical, wood must have some kind of protective finish coating. The finish might also be colored to add a decorative element to the instrument. On acoustic guitars, the finish must protect the instrument while not adding too much weight or damping.

There have been many solutions to the problem of making a guitar structure. Solid body guitar structures are generally so much stronger than they need to be that they can be shaped in many different ways. Indeed, there is a whole group of electric guitars that are intended to make a visual statement. Figure 3.1 shows a Jackson electric guitar with a stylized body. It doesn't need to be this shape in order to work – compare it to something more traditional like a Stratocaster or a Les Paul. However, it looks cool and that's important. There is an old saying that people hear first with their eyes. The shape of this guitar suggests the kind of music one might play on it. I have a hard time imagining someone playing a soft ballad on this guitar, though there is certainly no reason you couldn't. Scream-out heavy metal, however, is somehow easier to imagine.

It is perhaps helpful to think of the structure of acoustic guitars like a Haiku. The physical requirements are much more limiting and designers need to express their creativity within much harder boundaries. Partly for this reason, acoustic guitars tend to be more traditional in design.

Finally, guitars are mostly made from wood. Partly, this is by tradition; early luthiers simply didn't have many choices in materials and it was they who established what we now consider to be classic designs. It was they who taught

R.M. French, *Technology of the Guitar*, DOI 10.1007/978-1-4614-1921-1_3,
© Springer Science+Business Media New York 2012

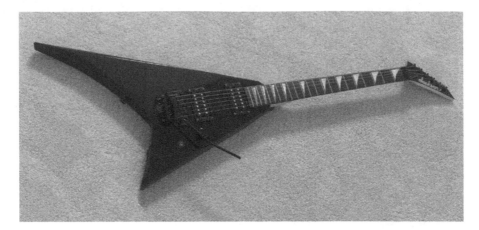

Fig. 3.1 A Stylized Jackson Electric Guitar (Wikimedia Commons, image is in the public domain)

Fig. 3.2 Ovation Acoustic Bass with Carbon Fiber Laminate Top and Bowl Back (Wikimedia Commons, image is in the public domain)

us how a guitar should look, how it should be made and how it should sound. That said, there is no reason that other materials can't be used. Fiber reinforced plastics have been used since the 1970s. Ovation Guitars is probably the most well-known manufacturer of composite instruments. Figure 3.2 shows an Ovation acoustic bass with the characteristic bowl back and array of small soundholes.

As instrument-quality wood is becoming more expensive, other materials are being explored. Martin is now using high pressure laminates to make durable guitars that sound good. Figure 3.3 shows an instrument with a body made using high pressure laminate (HPL) and a neck made of laminated wood. The HPL can be made with any design printed on it, so it offers wide aesthetic options. Some guitars use HPL printed to look like traditional wood. Others, like the one here use the top as a palette for elaborate graphics.

Figure 3.4 shows several partially completed necks made of laminated wood. This particular material is trade named Stratabond [1] and is akin to plywood except

Fig. 3.3 A Martin Acoustic Guitar made from High Pressure Laminate (Image courtesy of C.F. Martin Archives, http://www.martinguitar.com)

Fig. 3.4 Partially Completed Acoustic Guitar Necks Made of Laminated Wood (Image courtesy of C.F. Martin & Co., http://www.martinguitar.com)

that the grain is all going in the same direction. Traditional plywood is made from successive layers whose grain directions are perpendicular.

There are several potential advantages of a laminated neck over a more conventional solid wood one. These include:

- It is much less likely to have internal material flaws since all the laminations can be screened for cracks, knots and other imperfections before they are glued together. A solid wood neck may have internal flaws not visible from the outside.
- The glue seams are aligned with the centerline of the neck and can easily be arranged to be the equivalent of perfectly a quarter-sawn plank as shown by

the three examples in Fig. 3.4. These necks are very unlikely to warp or otherwise distort.

- There can be less environmental impact since individual laminations can be made from pieces that are otherwise too small to be useful.

3.1 String Forces

The most obvious forces applied to the structure of the guitar come from the strings. The strings are in tension, so there is a static force. However, since they vibrate, there is also a dynamic force. The static forces due to tension are much larger than the dynamic forces, which are usually ignored. The frequency expression for ideal strings can be re-arranged to solve for static tension.

$$T = 4f^2L^2\rho \tag{3.1}$$

Where f is the fundamental frequency, L is the string length and ρ is the mass per unit length. Table 3.1 shows string tensions provided by D'Addario for electric guitar strings of varying sizes. Guitar designers must assume that some players will choose heavy strings, so the structure must be able to withstand at least 81 kg (178.6 lb) of force. Again kg is being treated as a unit of force, even though this is not technically correct. Of course, temperature and humidity changes may temporarily increase this value. Thus, it might be reasonable to design a guitar for a maximum tension in the neighborhood of 95 kg (209 lb).

Note that string sizes and tensions depend on the type of instrument. Acoustic steel string guitars tend to use heavier strings than do electric guitars. On an electric guitar, the sound comes from the pickups and amplifier, and volume can be adjusted electronically. On an unamplified acoustic guitar, the sound comes solely from the kinetic energy of the strings. Consequently, it is difficult to get adequate volume from very light strings.

The sizes of strings in a set are generally identified by the diameter of the smallest string, the high E. Light electric guitar strings have 0.009 in or 0.010 in (0.229 mm or 0.254 mm) diameter high E strings. One hears these string sets referred to as 'nines' or 'tens'. Conversely, a light set of acoustic strings might have a high E string diameter of 0.012 in (0.305 mm). A heavy set of acoustic strings might have a high E string diameter of 0.014 in (0.356 mm). Table 3.2 presents string tensions for a range of acoustic string sets.

The maximum string tension for acoustic strings is more than 20% higher than that for electric guitar strings. This means that the structure of an acoustic guitar must be light enough and flexible enough to move in response to the motion of the vibrating strings while being able to withstand in-plane tension equivalent to the weight of a large man.

Finally, classical guitars use nylon strings that are significantly less dense than steel strings. The density of steel is 7850 kg/m³ while that of nylon is about 1120 kg/m².

Table 3.1 Static Tension of Selected D'Addario Electric Guitar Strings

Note	Dia (in)	Dia (mm)	Tension (Lb)	Tension (kg)
	Extra Super Light	EXL130		
E	0.008	0.20	10.4	4.72
B	0.010	0.25	9.1	4.13
G	0.015	0.38	12.9	5.85
D	0.021w	0.53	12.0	5.44
A	0.030w	0.76	14.0	6.35
E	0.038w	0.97	12.1	5.49
	Total		**70.5**	**32.0**
	Regular Light	EXL 110		
E	0.010	0.25	16.2	7.35
B	0.013	0.33	15.4	6.98
G	0.017	0.43	16.6	7.53
D	0.026w	0.66	18.4	8.34
A	0.036w	0.91	19.5	8.84
E	0.046w	1.17	17.5	7.94
	Total		**103.6**	**47.0**
	Jazz Medium	EJ22		
E	0.013	0.33	27.4	12.43
B	0.017	0.43	26.3	11.93
G	0.026w	0.66	32.8	14.88
D	0.036w	0.91	34.8	15.78
A	0.046w	1.17	31.1	14.1
E	0.056w	1.42	26.3	11.93
	Total		**178.7**	**81.05**

Note: w indicates wound strings

The result is that a steel E string with a diameter of 0.012 in (0.305 mm) has a mass per unit length of 0.573 gm/m. A nylon E string with a diameter of 0.028 in (0.711 mm) has a mass per unit length of 0.445 gm/m. Table 3.3 shows the published tensions for two popular sets of classical guitar strings.

Because of the lower string tension, classical guitars can be lighter than steel string guitars. Table 3.3 shows that a typical set of classical guitar strings has a maximum tension of about 40 kg (392 N or 88.2 lb). As a result, the braces can be much smaller and the tops can be thinner.

It's worth pausing at this point to try some sample calculations to see whether the published tension values match numbers predicted by Equation 3.1 above. Let's start by considering a 0.010 in (0.254 mm) steel E string. The data in the three tables here assume a string length of 25.5 in (647.7 mm), so we will use the same value. Figure 3.5 shows the calculation as performed using Mathcad. The predicted tension of 16.3 lb (7.40 kg) is very close to that shown in Table 3.1.

Using the same process, the predicted tension of a steel high E string with a diameter of 0.012 in is 23.78 lb (10.79 kg).

So far, all the calculations have been for unwound strings, however, some strings are made of steel core wires overwound with softer wire. Windings are applied only

Table 3.2 Static Tension of Selected D'Addario Acoustic Guitar Strings

Note	Dia (in)	Dia (mm)	Tension (Lb)	Tension (kg)
	Light	EJ 16		
E	0.012	0.30	23.3	10.57
B	0.016	0.41	23.3	10.57
G	0.024w	0.61	30.2	13.70
D	0.021w	0.81	30.5	13.83
A	0.030w	1.07	29.9	13.56
E	0.038w	1.35	26.0	13.15
	Total		**163.2**	**75.38**
	Medium	EJ 17		
E	0.013	0.33	27.4	12.43
B	0.017	0.43	26.3	11.93
G	0.026w	0.66	35.3	16.01
D	0.035w	0.89	36.8	16.69
A	0.045w	1.14	34.0	15.42
E	0.056w	1.42	29.0	13.15
	Total		**188.8**	**85.63**
	Heavy	EJ 18		
E	0.014	0.36	31.8	14.42
B	0.018	0.46	29.5	13.38
G	0.027w	0.69	38.4	17.41
D	0.039w	0.99	45.2	20.50
A	0.049w	1.24	40.0	18.14
E	0.059w	1.50	32.2	14.60
	Total		**217.1**	**98.45**

Table 3.3 Static Tension of Selected D'Addario Classical Guitar Strings

Note	Dia (in)	Dia (mm)	Tension (Lb)	Tension (kg)
	Normal Tension	EJ45		
E	0.028	0.711	15.3	6.94
B	0.0322	0.818-	11.6	5.26
G	0.0403	1.024	12.1	5.49
D	0.029w	0.74	15.6	7.08
A	0.035w	0.89	15.0	6.80
E	0.043w	1.09	14.0	6.35
	Total		**83.6**	**37.92**
	Hard Tension	EJ46		
E	0.0285	0.724	15.8	7.17
B	0.0327	0.831	12.0	5.44
G	0.0410	1.041	12.4	5.62
D	0.030w	0.76	16.3	7.39
A	0.036w	0.91	15.9	7.21
E	0.044w	1.12	14.5	6.58
	Total		**86.9**	**39.41**

Predicted Tension of a Steel Guitar String

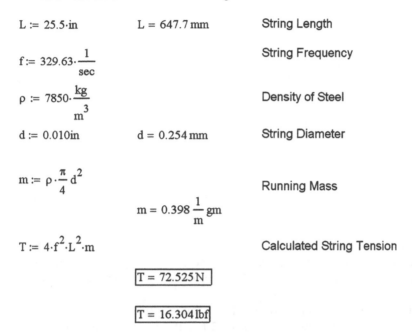

$L := 25.5 \cdot in$ $L = 647.7\,mm$ String Length

$f := 329.63 \cdot \dfrac{1}{sec}$ String Frequency

$\rho := 7850 \cdot \dfrac{kg}{m^3}$ Density of Steel

$d := 0.010\,in$ $d = 0.254\,mm$ String Diameter

$m := \rho \cdot \dfrac{\pi}{4} d^2$ Running Mass

$m = 0.398\,\dfrac{1}{m}\,gm$

$T := 4 \cdot f^2 \cdot L^2 \cdot m$ Calculated String Tension

$$\boxed{T = 72.525\,N}$$

$$\boxed{T = 16.304\,lbf}$$

Fig. 3.5 Calculation of String Tension Using Mathcad, Steel E String, d = 0.010 in.

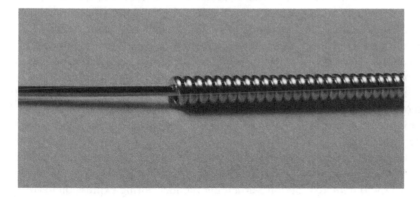

Fig. 3.6 Close Up of a Wound String

to the lower frequency strings and are used to increase the mass without increasing the diameter of the load carrying part of the string. Figure 3.6 shows an extreme close up of a wound string. Note that the core wire of this string, made by D'Addario, has a hexagonal cross-section. The resulting six corners help prevent slippage between the windings and the core.

Without windings, low pitch strings would need to have large diameters. While this would increase the mass, it would also greatly increase bending stiffness and

Fig. 3.7 A Taylor 12 String Acoustic Guitar (Image courtesy of Taylor Guitars, http://www.taylorguitars.com).

create inharmonicity problems. Music wire, defined by ASTM A228 [2], has a very high yield stress, so the core wire doesn't need to be large to withstand the tension required to bring a string to pitch. This is especially true if the string is mass loaded with windings.

The extreme case is bass strings. Bass scale lengths can exceed 34 in (864 mm) and the pitch of the low E string is 41.2 Hz. If this string was solid and made of music wire, the ideal string equation predicts that the diameter would have to be about 3.18 mm (0.125 in). This would be more of a steel rod than a string and would be completely unworkable. Rather, low E bass strings generally have more than one layer of windings so that they can be very heavy while still being flexible.

When using the ideal string equation with wound strings, it is necessary to calculate the mass per unit length (running mass). Because the string has more than one component and is not solid, the correct value can't be calculated directly from material density and cross sectional area. There are basically two practical methods. One is to weigh a known length of string and calculate running mass. The other is to calculate the running mass using the ideal string equation and a tabulated value for tension.

Take the example of a wound string from Table 3.2. The diameter of the A string is 0.045 in (1.14 mm) and the desired frequency is 110 Hz. Finally, the tension is 15.42 kg (34.0 lb or 151.2 N). Re-arranging the ideal string expression to solve for running mass gives

$$\rho = \frac{T}{4L^2 f^2} = 7.45 \frac{gm}{m} \tag{3.2}$$

While not currently common, some instruments have more than six strings and the structure must be reinforced to withstand the increased string tension. The most familiar of these instruments is the 12 string guitar as shown in Fig. 3.7. This is a 12 string Taylor GA3-12 and is outwardly quite similar to their six string GA3 model.

Fig. 3.8 A Taylor 12 String Acoustic Guitar Showing the Strings in Six Courses (Image courtesy of Taylor Guitars, http://www.taylorguitars.com)

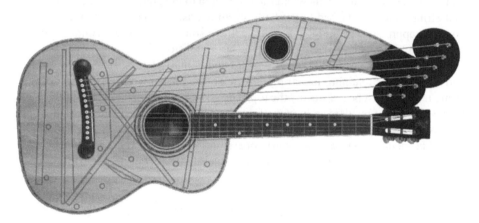

Fig. 3.9 A Modern Harp Guitar with the Bracing Pattern Superimposed (Image courtesy of Holloway Harp Guitars, http://www.hollowayharpguitars.com)

The combined tension of a representative set of medium strings for a 12 string acoustic guitar is 325 lb (147.4 kg).

On 12 string instruments, the strings are in six pairs as shown in Fig. 3.8. Thus, it is played essentially as if it were a six string instrument.

There are other, more unusual instruments with correspondingly unusual structural requirements. One that has been available in the United States since at least the beginning of the 20th century is the harp guitar. Indeed, one of Orville Gibson's first products was an archtop harp guitar. Figure 3.9 shows a modern, flat top harp guitar with six additional strings that can be plucked, but not fretted. The internal bracing pattern has been superimposed over the image of the instrument (unfortunately perhaps for guitar-playing engineers, these instruments are not sold with the bracing pattern printed on the top).

3.2 Body Structure of the Acoustic Guitar

For an acoustic guitar to make sound, the top must be able to move in response to string vibration. Of course, the top must also be able to resist the tensile forces of the string. Thus, designing the structure of an acoustic guitar body is an exercise in balancing the two competing requirements of being strong, yet flexible. The traditional wood acoustic guitar body is made with nearly flat top and back plates joined by bent sides. Typically, the top plate is made from a species of conifer such as spruce, cedar or redwood. The most popular material has been spruce, either Sitka spruce or Engelmann spruce, though the supply of high quality tops is limited and some other species are being used.

The primary load on the top of an acoustic guitar is string tension, and the effect on the top depends on how the strings are mounted to the guitar. There are two ways that strings can be attached to the body of an acoustic guitar. The first, and most common approach, is a bridge that is fixed to the soundboard. The second approach is a floating bridge that is held in place by string tension.

The fixed bridge is glued to the top plate and contains both a slot for the saddle and a means for retaining the strings. Because the bridge is glued to the top, the string tension must be resisted by the top. The strings are not in the plane of the top, so the equivalent result is an in-plane force on the top combined with a moment as shown in Fig. 3.10.

Unlike metal, wood creeps at room temperature. That means wood can permanently change shape (plastically deform), even under loads far lower than required to cause an

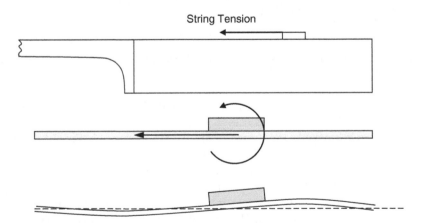

String Tension

Fig. 3.10 Effect of String Tension on a Guitar top

Fig. 3.11   An Acoustic Guitar with Top Deformed By String Forces

immediate failure. Conversely, metal does not plastically deform unless the stress exceeds an elastic limit known as yield stress or unless it is at very high temperature.

Since wood is subject to creep, guitars can distort over time. It is not uncommon for an instrument that looks fine when it leaves the factory to develop permanent distortions over time. Figure 3.11 shows an extreme case of soundboard distortion in an acoustic guitar. The load-bearing structure of this instrument is fairly light, presumably in an effort to improve sound quality. Even though the top was undistorted when built, it eventually developed severe distortion around the bridge.

For the luthier, the tendency of wood to creep means the structure must be designed for the long term. Lightly-built instruments tend to be more responsive to the string vibrations and this sensitivity is preferred by some players. However, the reduced stiffness can allow serious structural problems to appear over time. As with most other products, the guitar designer's role requires balancing opposing needs.

3.2.1   Bridges

On almost all acoustic guitars, the bridge serves both as a mounting point for the saddle, which forms the end of the vibrating string, and as the structure that anchors one end of the string, allowing it to transfer the tension load to the body. Figure 3.12 shows a close up of an acoustic bridge.

In most cases, acoustic bridges are simply glued to the top. The shear strength of even weak glue is enough to resist the pull of the strings. Say a set of acoustic strings has a combined tension of 625 N (140.5 lb) and the bridge has an approximate gluing area of 4600 mm^2 (7.13 in^2). The resulting shear stress is 0.135 MPa (19.64 psi), far below the shear strength of even weak glue.

The actual stress between the top and the bridge depends on how the strings are fixed to the bridge and the top. Steel strings have ball ends at the bridge and are generally fixed to the bridge using bridge pins as shown in Fig. 3.12. The cross section is shown in Fig. 3.13. Since the ball end is in contact with the reinforcing

Fig. 3.12 An Acoustic Bridge with Saddle (Image courtesy of Hagstrom Guitars, http://www. hagstromguitars.com)

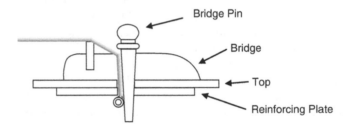

Fig. 3.13 Cross Section of an Acoustic Bridge with Saddle and Bridge Pin

plate (often called a bridge plate), it is able to transmit force directly from the string to the top. This doesn't reduce the shear force, but adds a normal force pushing the bridge against the body. Bridge pins are usually made of wood or plastic. Their purpose is just to keep the ball ends from pulling through the holes in the top.

Almost all classical guitars and a few steel string acoustic guitars anchor the strings directly to the bridge. In classical guitars, this is done by tying the strings to a tie block behind the saddle; classical strings do not generally have ball ends, so it is necessary to tie them in order to keep them from slipping. Figure 3.14 shows the saddle and tie block on a classical guitar bridge.

There is one type of bridge not directly fixed to the top. It is called a floating bridge and is held to the body only by the vertical component of string tension. This method, universally employed on violins, is often used on archtop guitars. It offers the advantage of being able to move the bridge to adjust intonation as needed, but the bridge will fall off if the all the strings are removed. Figure 3.15 shows a Manhattan model archtop jazz guitar by Bob Benedetto with a floating bridge.

Fig. 3.14 A Classical guitar Bridge Showing the Tie Block (Image by the author, reproduced courtesy of Prof. Eugene Coyle)

Fig. 3.15 A Benedetto Manhattan Archtop Guitar with a Floating Bridge (Image courtesy of Bob Benedetto, http://www.benedettoguitars.com)

Fig. 3.16 Forces Acting on a
Floating Bridge

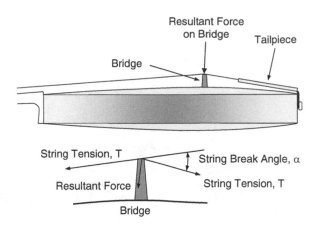

A less-obvious effect of using a floating bridge is that it changes the forces on the guitar body. Since the strings pass over the bridge and are secured to a tailpiece separate from it, the primary force on the bridge is a vertical one that pushes it against the top. Figure 3.16 shows the forces acting on a floating bridge on an archtop guitar.

3.2.2   Bracing Patterns

Almost all acoustic guitars have tops consisting of thin plates with bracing bars glued to the inside surface. The arrangement of these bars is called the bracing pattern. Steel string acoustic guitars typically have heavier braces than classical guitars since the string tension is higher. The most common bracing pattern for flat top, steel string acoustic guitars is called X-bracing. Fig. 3.17 shows a cutaway view of a Martin acoustic guitar with X-bracing.

The two heavy braces cross right below the soundhole and there is a reinforcing plate between them that spreads the loads from the bridge. There are also lighter braces adjoining the X braces and a cross brace above the soundhole. While there is no standard X brace pattern, this one is quite representative.

The backs of steel string guitars are also typically braced, though with a simpler pattern as shown in Fig. 3.18. This back is for a Taylor guitar, but is very similar to the back on the cutaway instrument in the previous figure. The center strip is a made from thin wood cut so that the grain runs perpendicular to the center line of the top. This reinforcement helps keep the center seam from separating over time. The parallel cross braces are roughly evenly spaced along the back. This parallel bracing pattern is often called ladder bracing and was formerly used for tops on guitars and lutes.

It is important to note that most acoustic guitars have domed tops and backs. The curvature is generally low and expressed in radius of curvature. Rather than being

Fig. 3.17 Cutaway View of an Acoustic Guitar Showing X-Bracing (Image courtesy of C.-F. Martin Archives, http://www.martinguitar.com)

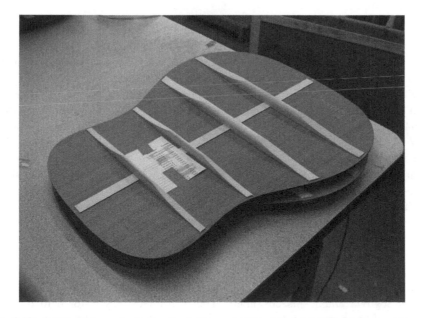

Fig. 3.18 A Braced Acoustic Guitar Back (Image by the author, reproduced here courtesy of Taylor Guitars, http://www.taylorguitars.com)

Fig. 3.19 Interior of an 1863 Torres Guitar Restored by Richard Bruné (Images courtesy of Richard Bruné, http://www.rebrune.com)

flat, the ideally domed top or back can be thought of as having been cut from the surface of a sphere. The curved shape is fixed by shaping the braces and gluing them while the plate is held in a fixture with a concave face. Pressure is usually applied either with thin flexible rods (called go bars) or with a vacuum.

Different designs use different radii for tops and backs. However, top radii are often in the range of 28 ft – 40 ft (8.53 m – 12.2 m). Back radii are generally shorter so that backs have more curvature. Typical values are in the range of 15 ft – 20 ft (4.57 m – 6.10 m).

It is worth pausing to consider the role of bracing in the design of acoustic guitars. The obvious purpose is to add stiffness to the top plate, which serves both as a structural element and as a solid surface that can radiate sound. However, it is certainly possible to make the top thick enough that bracing isn't required.

Figure 3.19 shows interior of the soundboard in a historically important instrument, a very rare Torres guitar from 1863 [3]. It was designed to be inexpensive and affordable for players of modest means. This particular instrument has been carefully restored by Richard Bruné, one of the most accomplished luthiers now working and a historian of the early guitar. He has the rare distinction of having Andres Segovia among his customers. The instrument was restored to playing condition after having been found in a state of near wreckage. The picture here is of the instrument part way through the restoration process.

Fig. 3.20 Richard Bruné Playing the Restored 1863 Torres Guitar (Image by the author, reproduced courtesy of Richard Bruné, http://www.rebrune.com)

Bruné observes that the lack of bracing might be an insight into how Torres viewed its role. That is, the top is there for structural reasons and the bracing serves to tune the response of the top to improve the sound. For a very low cost instrument, he apparently considered the bracing to be unnecessary. It should be noted that this is a very small instrument, akin to what might later have been called a parlor guitar. Figure 3.20 shows Bruné playing the restored instrument and clearly showing its small size.

The most common bracing pattern for classical guitars is fan bracing, the pattern used by Torres starting in the mid 19th century. Figure 3.21 shows an example of fan bracing in a 1952 guitar by Herman Hauser Sr. This instrument was in the process of being restored by Richard Bruné and the back had been removed. The seven braces are clearly seen along with three cross braces at the waist and upper bout. Also clear are the doubler under the bridge and the two diagonal braces at the lower ends of the fan braces.

Figure 3.22 shows a classical guitar under construction. There are seven fan braces and two cross braces, but no bridge plate. Note that the cross braces extend to the sides and are secured by cleats glues to the sides. The notches in the kerfing where the back will be glued are where the back braces will join the sides.

Luthiers have experimented with many different bracing patterns for classical guitars and some of the results have been quite interesting. One well-known design was developed by Dr. Michael Kasha (a distinguished physical chemist) and

Fig. 3.21 A Fan Braced
Classical Guitar Made by
Herman Hauser Sr in 1952
(Image by the author,
reproduced courtesy of
Richard Bruné, http://www.
rebrune.com)

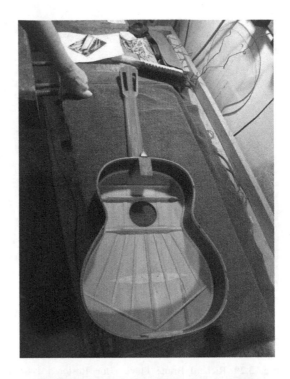

Fig. 3.22 A Classical Guitar
with Light Seven Bar Fan
Bracing (Image courtesy of
Sophie Karolidis)

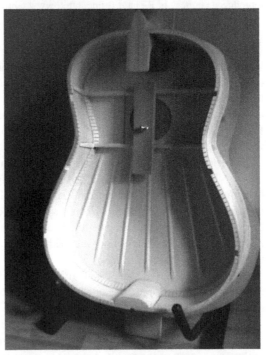

U.S. Patent   March 21, 1978   Sheet 1 of 4   **4,079,654**

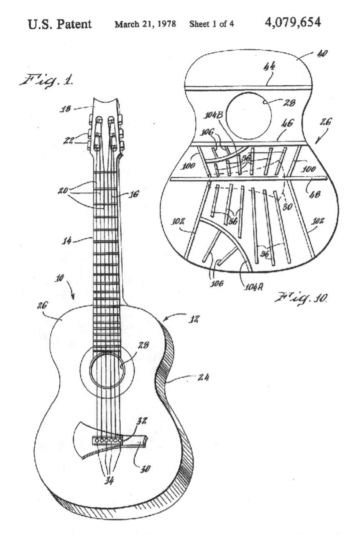

Fig. 3.23 Patent Drawings Showing the Characteristic Kasha Bridge and Bracing Pattern

implemented by luthier, Richard Schneider. Kasha was awarded a patent for an early version of his design in 1969 [4] and patented a more refined version in 1978 [5]. Figure 3.23 shows a patent drawing of a guitar and the bracing pattern.

There are several basic ideas underlying this design. One is that bracing should not cross the centerline of the bridge. Rather, the torque created by the strings acting at the top of the bridge is distributed to the rest of the structure by a torsion bar running under the bridge across the lower bout. Another key idea is that different areas of the soundboard are tuned to have specific resonant frequencies. The 1978 patent shows a number of different bracing patterns intended to create a region with a high resonant frequency on the treble side of the soundboard and a region with a

Fig. 3.24 A Gibson MK-81
Guitar with Kasha –
Schneider Structure
(Wikimedia Commons,
image is in the public domain)

low resonant frequency on the bass side. Some instruments also have an offset soundhole – usually shifted to the treble side of the upper bout.

A characteristic external feature of Kasha – Schneider guitars is the asymmetric bridge. It is intended to work with the bracing in allowing different sections of the soundboard to have different resonant frequencies. Some versions of this bridge are significantly cut away between the 3rd and 4th strings. This essentially divides the bridge into two relatively rigid sections joined by a short section with greatly reduced stiffness. Figure 3.24 shows a Gibson MK-81 acoustic guitar, a design based on the work of Kasha and Schneider and produced in limited quantities.

A more recent development in acoustic guitars – usually classical guitars - is lattice bracing. Bracing serves to distribute stiffness across the soundboard and a natural extension of this idea is to replace a relatively small number of braces with a network of smaller, interlinked ones. Figure 3.25 shows a lattice braced top by Chris Pantazelos. This top is for a non-standard seven string instrument, but is representative of tops made for more conventional six string instruments.

Fig. 3.25 A Lattice Braced Top (Image courtesy of Chris Pantazelos, http://www.spartan instruments.com)

Figure 3.26 shows an instrument by Steve Connor with lattice bracing and an additional sound hole that faces the player. This bracing pattern is related to a traditional five bar fan brace, but add lateral braces to form a lattice.

Generally, instrument builders try to make tops with the required stiffness and as light as possible. A logical extension of the lattice bracing approach is to reinforce the braces with a light stiff material like carbon fiber. Some luthiers glue thin strips of unidirectional carbon fiber (also called graphite) to the tops of braces and this approach can be used for lattice braces.

Figure 3.27 shows such a top made by Greg Smallman, an accomplished Australian luthier known for innovative designs. Note that the braces taper from full height at the center of the soundboard to almost zero height at the ends. The stiffness of a beam is proportional to the cube of its height, so tapering like this is a very effective way to tailor the stiffness of the top.

Some builders have experimented with building a portion of the load carrying structure into a frame that also supports the top at the lower bout. The idea is to lower the stiffness required from the top so that it can be lighter and more responsive to the motion of the strings. Figure 3.28 shows an experimental guitar by Dave Schramm and inspired by the work of the late Jim Norris that combines lattice bracing with a heavy load-carrying frame around the edge of the lower bout.

Figure 3.29 shows a top view of the same instrument before the top is glued on. There is an oval relief in the frame so that the lattice braced top is free to vibrate. There is no similar feature in the upper bout, so the top is essentially rigid in that area.

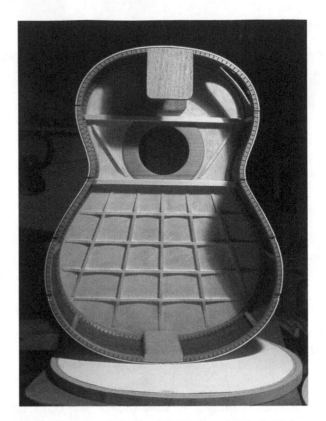

Fig. 3.26 A Lattice Braced Classical Guitar with Side Sound Hole (Image courtesy of Stephan Connor, http://www.connorguitars.com)

Fig. 3.27 A Lattice Braced Top by Greg Smallman with Graphite Reinforcement (Wikimedia Commons, image is in the public domain)

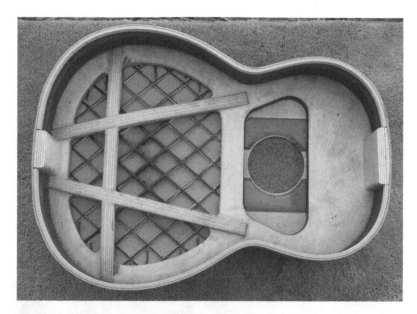

Fig. 3.28 An Experimental Lattice Braced Guitar with a Heavy Load Bearing Frame (Image courtesy of Dave Schramm, http://www.schrammguitars.com)

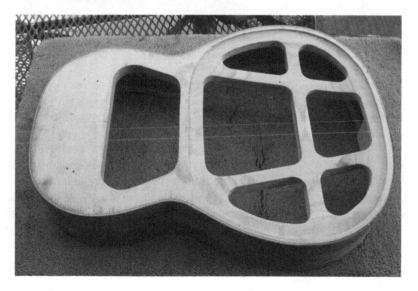

Fig. 3.29 Frame of Guitar before Lattice Braced Top is attached Frame (Image courtesy of Dave Schramm, http://www.schrammguitars.com)

Another approach to reducing the load on the top is to add reinforcing rods in the body, often running between the tail block and the neck mounting structure. Figure 3.30 shows a partially completed body made by Gary Southwell with two sets of graphite reinforcing rods. The rods are intended to distribute loads from string tension so that lower loads are imposed on the soundboard.

Fig. 3.30 A Partially
Completed Body
Incorporating Carbon Fiber
Rods (Image courtesy of Gary
Southwell, http://www.
southwellguitars.co.uk)

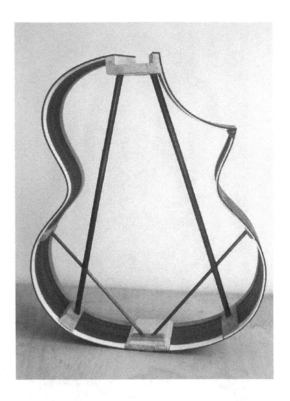

It has been known for some time that sandwich structures can be light, stiff and strong. Structurally speaking, a sandwich consists of two strong outer layers surrounding a light core. A familiar example is a surf board made from a light foam core with fiberglass skin. A typical surf board is very light – on the order of 5 kg (11 lb) – yet strong enough to support a person and to withstand being tossed around in turbulent waves. Sandwich structures are very common in aerospace applications where strength, stiffness and light weight are often required.

Sandwich structures would seem to be well suited to guitars, and some builders are using them for guitar tops. A popular core material is Nomex honeycomb, a light, stiff paper-like material joined to form hexagonal cells. It is widely used in aerospace applications. Figure 3.31 shows an instrument made by Randy Reynolds with a honeycomb sandwich top. It is being illuminated from the inside so that the interior structure is visible.

The top consists of two very thin spruce plates with a honeycomb core. Note the slightly darker hourglass shaped area under the center of the bridge and extending to the area surrounding the soundhole. This is an area in which the honeycomb core has been replaced by thin plywood to increase the strength. Note also that there are braces on the inside surface of the top. These are arranged in a pattern similar to those developed by Michael Kasha.

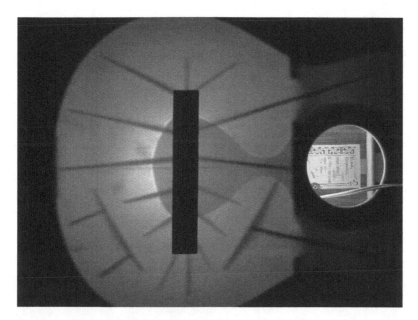

Fig. 3.31 A Guitar with a Sandwich Top Illuminated from the Inside (Image courtesy of Randy Reynolds, http://www.reynoldsguitars.com)

One potential problem with sandwich structures is bonding the layers together. It is easy to add unacceptable weigh to an otherwise light structure by using too much glue to bond the layers together. Reynolds avoids the problem by spreading a thin layer of epoxy on a sheet of glass and then pushing the honeycomb into it, wetting only the edges of the Nomex cell walls. Also, a layer of glue that is soft, even when cured, has the potential to add damping to the top.

3.2.3 Bracing Stiffness

One topic that seems to be missing from many works on guitar making is the relationship between the cross-sectional shape of braces and their stiffness. This is a topic generally covered in engineering classes on strength of materials. Without getting into too much detail, it is approximately correct to treat braces as if they were beams.

Beam stiffness is due to two characteristics: material and geometry. The material stiffness is described by its elastic modulus. Materials with high elastic modulus are stiffer than materials with low elastic modulus. Because elastic modulus is defined as the ratio of stress to strain, it has units of pressure (force per unit area). Sitka spruce, a species often used for tops and braces, has an elastic modulus of about 8.2 GPa (about 1,200,000 psi). Since wood is fibrous, there are actually elastic moduli that describe material stiffness in the different directions. The number here is measured along the grain.

Table 3.4 Area Moments of Inertia and Section Moduli for Common Shapes

Rectangle	h / b	$I = \dfrac{bh^3}{12}$	$s = \dfrac{h^2}{12} = 0.0833h^2$
Triangle	h / b	$I = \dfrac{bh^3}{36}$	$s = \dfrac{h^2}{18} = 0.0556h^2$
Ellipse	h / b	$I = \dfrac{b}{2}h^3\left(\dfrac{\pi}{8} - \dfrac{8}{9\pi}\right)$	$s = h^2\left[\dfrac{1}{4} - \dfrac{16}{9\pi^2}\right] = 0.0699h^2$

Of perhaps more interest here is the relationship between stiffness and cross-sectional shape. The number that describes this quantity is the area moment of inertia, identified as I. Different shapes have different moments of inertia and the expressions are usually tabulated in textbooks and engineering reference books. Deriving these expressions requires advanced math and is beyond the scope of this book. Rather, Table 3.4 lists properties for several cross sectional shapes useful to guitar designers and makers. Note that the area moments of inertia are calculated about the centroids of the respective shapes.

Even more useful for guitar makers than area moment of inertia is section modulus, s. The section modulus is simply area moment of inertia divided by the cross-sectional area. Assuming the material doesn't change, the section modulus is stiffness per unit weight. Based on section modulus, the rectangle is the stiffest shape per unit weight.

This is not the whole story since the braces are glued to the soundboard, but it is enough to make the point that triangular or elliptical braces are not as efficient, as defined by stiffness per unit weight, as rectangular ones. Using this definition, a brace with a T-shaped cross section would be even more efficient. However, the builder would have to take care to ensure that there was enough gluing area to connect it to the top.

3.3 Glued Joints

It may seem picky to pay attention to glued joints, but their designs as well as the properties of the glues themselves are critical to the design of a guitar. Indeed, a weak glue joint can ruin an otherwise good instrument. The most important thing to

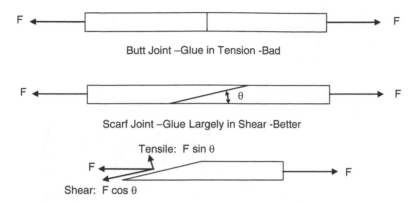

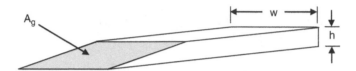

Fig. 3.32 Example of Bad and Good Design of a Glued Joint

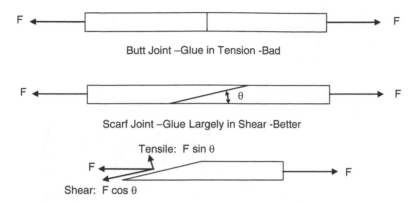

Fig. 3.33 Gluing Area of a Scarf Joint

know about designing string glue joints is that glue is only effective in shear. It is of little use in compression or tension. Figure 3.32 shows a simple example.

It might be helpful to have some numbers to make the example a little less abstract. Say that the angle of the scarf joint is 20°, the force is 450 N (101 lb) and the cross section of the board being joined is 15 mm x 45 mm (0.591in x 1.772in). For the butt joint, the tensile stress would be 667 kPa (96.7 psi).

$$F_\tau = F sin\theta = 450N\ sin(20°) = 450N \times 0.342 = 153.9N = 34.6lb$$

The tensile force on the glued surface is only about a third of the force acting on the board. To calculate the tensile stress, σ, we simply divide the tensile force by the gluing area, A_g, as shown in Fig. 3.33.

The resulting tensile stress is

$$\sigma = \frac{F_\tau}{A_g} = \frac{Fsin\theta}{wh/sin\theta} = \frac{F}{wh}sin^2\theta = \frac{450N}{0.045m \times 0.015m}\ 0.1170 = 78,000Pa$$

$$= 11.3\frac{lb}{in^2}$$

This tensile stress is extremely low and just about any adhesive can withstand it easily. The shear stress on the joint is calculated in basically the same way, though with the force component parallel to the joint. The shear stress is

Fig. 3.34 A Finger Joint in a Neck Blank (Image by the author, reproduced courtesy of Taylor Guitars, http://www.taylorguitars.com)

$$\tau = \frac{F_s}{A_g} = \frac{Fcos\theta}{wh/sin\theta} = \frac{Fsin\theta cos\theta}{wh} = \frac{450N \times 0.342 \times 0.940}{0.045m \times 0.015m} = 214,000Pa = 31.1\frac{lb}{in^2}$$

The ratio of tensile stress to shear stress in a scarf joint is a string function of the angle. If $\theta = 15°$, then $\sigma = 44,700$ Pa = 6.48 psi and $\tau = 167,000$ Pa = 24.2 psi.

Failure stress for high quality glues in shear is often in the neighborhood of 3,000 psi (20.7×10^6 Pa or 20.7 MPa), so these stresses are both extremely low. However, the numbers reported for shear strength of glue are collected in carefully controlled experiments, usually according to ASTM standard D-905 [6]. In practice, there are many factors that might affect the strength glued joints, so there need to be substantial margins of strength.

A nice example of a glue joint with a large amount of shear area in a small space is the finger joint that has been used on Taylor guitars as shown in Fig. 3.34. It is the same type of joint sometimes used to join segments of lumber to make longer boards. Taylor cuts the finger joint so that the headstock is correctly angled. Taylor guitars use a headstock angle of 16° [7].

As Taylor has continued to refine their designs, they have developed a new neck joint they believe is superior to the finger joint. Figure 3.35 shows the progression from assembly to the completed neck. It is a variation on the traditional scarf joint with curved gluing faces. Making this joint requires being able to mill curved surfaces very precisely. Taylor's heavy use of CNC milling equipment clearly helps in bringing a design feature like this to production.

Temperature greatly affects the strength of glued joints. Glue generally weakens as temperature increases. Yellow wood glue can lose more than half its strength at 150 °F (65.6 °C). While this temperature would be unbearable in any inhabited building, it is certainly possible for the interior of a closed car to reach this

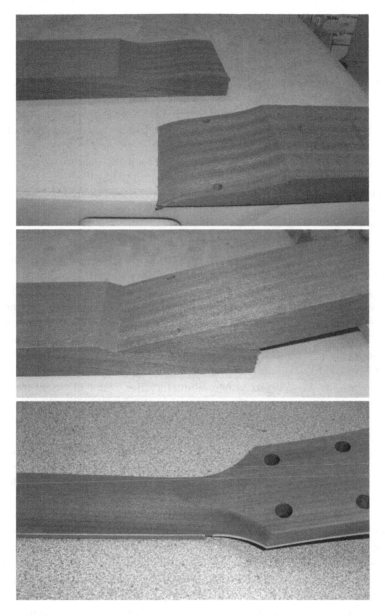

Fig. 3.35 A Curved Scarf Joint for Joining Headstock to Neck (Images courtesy of Taylor Guitars, http://www.taylorguitars.com)

temperature on a hot, sunny day. It is reasonable for a luthier to assume that an instrument might be left in a car under these circumstances.

Also, the strength of a glued joint is a strong function of the thickness of the glue layer. Glue usually works best when it forms a thin, uniform layer under shear stress.

If the glue layer is too thick, the joint can be weak. Conversely, if the wet joint is clamped together so tightly that all the glue is squeezed out, there may not be enough glue remaining to bear the applied loads.

As important as the design of glued joints are the material properties of the glue. Glue has been used for many thousands of years, with the earliest known examples dating from the middle stone age – approximately 70,000 years ago [8]. However, it has only been since the industrial revolution that man-made glues have been widely used. Before then, glue was made from natural materials such as egg, animal hides and hooves.

For example, the great violins, made in the 17th and 18th centuries by Stradivari, Guarneri and Amati, were all assembled using hot hide glue. Indeed, most violins are still made that way. The industrial revolution and rise of industrial chemistry led to the development of man-made glues with vastly superior mechanical properties to natural glues.

The glues used in modern instrument making fall roughly into four categories: hide glue, Polyvinyl Acetate (PVA), cyanoacrylate (CA) and two part or catalyzed glues. It is worth briefly examining the characteristics of each.

Hide glue is by far the oldest and most traditional adhesive available to luthiers. It is made by processing animal hides and is usually provided in flakes or granules. To make the glue ready for use, it is soaked in water to re-hydrate it and then heated to 60 °C (140 °F). As the glue cools, it gels quickly. Gelled hide glue is not very strong, so it is important to fit the pieces together quickly. The glue needs to be at least 50 °C (120 °F) when the pieces are brought together. It is not uncommon for luthiers to heat the wood before gluing. Some factories even had gluing rooms that were kept very warm to extend the working time. Without heating the parts or the room, working time for hot hide glue is quite short.

Hide glue has some notable advantages. If heat and humidity are controlled, it is as strong as some other types of glue. It also reversible – a joint can usually be separated with heat, sometimes even just a hot knife. Hide glue will stick to itself, so joints that have been separated can be re-glued easily. Finally, hide glue does not creep under load; a joint formed with hide glue and placed under a steady load will not move. The fact that the dried glue is brittle allows joints to be opened sometimes by simply tapping them with a knife. Violins, when well made and maintained, can last for centuries. An important reason for this is that the hide glue allows them to be disassembled for repairs when necessary. Additionally, some builders maintain that the low damping of the hardened hide glue contributes to the tonal quality of the completed instrument.

The disadvantages of hide glue are significant and there are some obvious reasons why it is not more widely used. Key among these is its inconvenience. Wood glue can simply be squeezed out of a bottle when necessary, but hide glue has to be prepared and heated. Liquid hide glue is available, but many luthiers distrust it, maintaining that it is not as strong as hot hide glue. Hot hide glue also loses strength quickly when temperature or humidity increases. While this allows easy disassembly of glued joints, it makes the completed instrument less durable in less than ideal conditions.

Perhaps the glue most familiar to modern luthiers is polyvinyl acetate (PVA) sold as a white general purpose glue (e.g. Elmer's) or, in a modified form, as a yellow wood glue (e.g. Titebond). Polyvinyl acetate is a rubbery synthetic polymer discovered in 1912 by the German chemist, Dr. Fritz Klatte. PVA is a water-based polymer that dries by evaporation. The load-bearing joint is formed as the glue soaks into the wood and the water evaporates. For obvious reasons, it works only on porous surfaces.

The original PVA was the white glue used by school children the world over. It is non-toxic, storable and easy to use. However, it is soft, even when dry, and sensitive to both heat and humidity. Yellow PVA, sold as wood glue, is harder when dry and more resistant to moisture. It is very widely used by individual builders and by large manufacturers. Newer formulations are even suitable for exterior use where water resistance is critical.

Cyanoacrylate (CA) is often called super glue, from the name of an early commercial brand. It is sold in different viscosities to suit a wide range of applications. They are widely used by luthiers, but usually, not in critical areas such as the neck joint.

The thinnest (lowest viscosity) version soaks readily into even tightly fitted joints and can be used with sawdust as a filler. An unwanted gap or void can be simply packed with fine wood dust and then wetted with thin CA. When it hardens, the result is a hard filler than can be sanded and finished. Higher viscosity formulations can be used when surfaces do not fit together perfectly or when slightly longer assembly times are required. Dried CA is brittle and can fail if subjected to impact loads.

The final category is two-part glues or catalyzed adhesives. This includes two-part epoxy and catalyzed polyester. Let's start with epoxy.

Epoxy is the name for a range of two-part polymer adhesives made from an epoxide resin and a polyamine hardener. They combine to form a material called polyepoxide. Epoxy is called a copolymer because it is formed from two different chemicals. When the resin is mixed with the hardener (often in roughly equal proportions by volume), they form a strong, cross-linked polymer – basically a hard plastic.

It is not very common for guitars to be made with epoxy as a structural adhesive, but it is often used for secondary purposes like gluing inlays. It is sometimes tinted to match the surrounding wood so that gaps between the inlay and the pocket are less visible. Two part epoxies are also sometimes used for finish. For example, Z-poxy finishing resin is a thin two part epoxy that flows nicely and can be used to form a shiny, durable finish.

Catalyzed polyester is also a two-part system that cures to forms a strong, clear polymer. However, the resin forms a large majority of the compound and it is combined with a small amount of a catalyst that initiates a reaction that cross links the polymer chains, converting the liquid resin to a solid. Polyester is commonly used as the matrix for composite materials like fiberglass. It is not generally used as an adhesive for guitars, but is very widely used commercially for finishes. There are safety issues that generally limit the use of polyester to manufacturing operations.

One other version of catalyzed polyester uses a catalyst that responds to ultraviolet light. The polyester/UV catalyst mixture is applied to the instrument (usually sprayed) and the instrument is then placed in a box where it is subjected to intense ultraviolet light. In a period usually on the order of a few minutes, the polyester hardens and can then be sanded and polished to a gloss finish.

3.4 Necks

The neck is a distinct structure from the body on most guitars. Structurally speaking, it can be treated as a beam cantilevered from the body on one end and free on the other end. The neck has to both withstand the tension of the strings without distorting too much and anchor one end of the strings. In most guitars, the neck is made separately from the body and is attached later, usually late in the build process.

There are basically three ways that necks are attached: bolted, glued and integral or neck through construction. Bolted necks originally appeared almost exclusively on solid body electric guitars, but are now common on acoustic guitars as well. This method was first widely used on the Fender Telecaster and Stratocaster, but has been adopted by other manufacturers. Figure 3.36 shows the back of a Stratocaster

Fig. 3.36 A Solid Body Electric Guitar with a Bolt On Neck

Fig. 3.37 A Simple Bolted Neck Joint for and Acoustic Guitar (Image courtesy of Stewart MacDonald, http://www.stemac.com)

with a bolt on neck. The rectangular plate is intended to distribute the force of the screw heads so that the body is not damaged. It's worth noting that most bolt on necks use screws rather than bolts. The typical neck screw is a 1 inch #8 oval head sheet metal screw.

Acoustic guitars are increasingly using bolted necks. The obvious advantage of a bolted joint is that the instrument can easily be taken apart when necessary. It is not uncommon for guitars to change shape slightly as they age and sometimes the angle of the neck with respect to the body changes enough to cause problems. When this happens, the neck must be reset at the correct angle. If the neck is glued on, the joint must be somehow disassembled without damaging the instrument.

The simplest bolted acoustic neck joints are extensions of traditional glued joints. Figure 3.37 shows an acoustic neck joint that uses a mortise and tenon for alignment and bolts to hold the neck in place. Note the three holes in the neck block that will eventually be inside the body of the instrument. The lower two holes are for the bolts – the nuts are tightened from the inside – and the upper one is for adjusting the truss rod.

Taylor Guitars is probably the most well-known of the companies using bolted neck joints and their mounting system may be the best developed. Their system uses three bolts, two for the heel and one through the bottom of the fretboard that overhangs the body. Figure 3.38 shows a stack of Taylor necks with the threaded inserts installed. All three bolts are inserted from inside the body.

Figure 3.39 shows the body of a Taylor acoustic guitar ready for the neck to be attached. Note that this is a left hand guitar. Also, there is a small magnetic pickup

Fig. 3.38 Taylor Necks with Threaded Fittings (Image by the author, reproduced courtesy of Taylor Guitars, http://www.taylorguitars.com)

Fig. 3.39 Body of a Taylor Acoustic Guitar Showing Neck Mounting Pockets (Image by the author, reproduced courtesy of Taylor Guitars, http://www.taylorguitars.com)

Fig. 3.40 A Selection
of Neck and Body Shims
for Taylor Guitars (Image
by the author, reproduced
courtesy of Taylor Guitars,
http://www.taylorguitars.com)

inset between the upper neck pocket and the soundhole. There are clearly three holes in the body to match the three in the neck.

Less obvious in this picture is the fact that the front and top pockets on the body are designed to accept specially made shims. Indeed, the instrument cannot be assembled without them. During any production process, dimensional errors can occur, and they tend to build up as the part moves from one assembly process to the next. Even though Taylor's build process is precise and well-controlled, there is some dimensional variation in the resulting necks and bodies. A primary purpose for the shims is to correct for these variations.

During assembly, the neck and body are placed in a fixture to determine which shims are needed in order to correctly align one to the other. Then, the appropriate shims are selected from a rack (Fig. 3.40) and inserted into the body pockets and the neck is installed. This system allows correct alignment quickly and reliably, but also makes the process of resetting the neck on an older guitar simple. The neck is removed from the body and they are placed back in the assembly jig. Then, new shims are selected and the instrument is reassembled.

Acoustic guitars have traditionally used glued neck joints. Steel string acoustics typically used a tapered dovetail joint as shown in Fig. 3.41. This type of joint is strong and has proven itself over time, but it must be made accurately because it

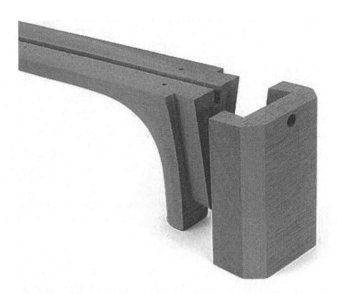

Fig. 3.41 A Tapered Dovetail Joint for an Acoustic Guitar (Image courtesy of Stewart MacDonald, http://www.stewmac.com)

makes the neck alignment difficult to adjust. For example, changing the neck angle might require inserting shims in the angled surface between the neck and the mounting block. Even then, there might be might be an unacceptable gap between the heel and the body.

Electric guitars also sometimes use glued on necks, though they are often called set necks. Probably the most popular guitar with a set neck is the Gibson Les Paul. The neck is at a small angle to the body in order to make the instrument more comfortable to play, and this must be accounted for in the neck joint. Figure 3.42 shows a Les Paul replica under construction. The neck has not yet been carved to the final shape, but the joint is complete. Note that the neck structure extends all the way back to the first pickup pocket.

A more recent set neck design is from the Parker Fly, an innovative design first introduced in the early 1990s. The designer, Ken Parker, sought to make a light, responsive instrument and used light woods with a composite exoskeleton. Figure 3.43 shows the highly contoured spruce body with the milled section ready to accept the neck.

Figure 3.44 shows the intersection of the neck and body on the completed instrument. The heavy contouring and the composite covering combine to form a flowing, almost liquid looking shape that is at once light, strong and comfortable to play.

The last type of neck joint is called a neck through body (or, in shorthand, a neck through) design. In this design, the block from which the neck is carved extends all the way through the body so that the neck and body form one continuous structure. Neck through designs typically require complex machining and are not common

Fig. 3.42 The Neck Joint in a Partially Completed Les Paul Reproduction (Image courtesy of http://www.mylespaul.com)

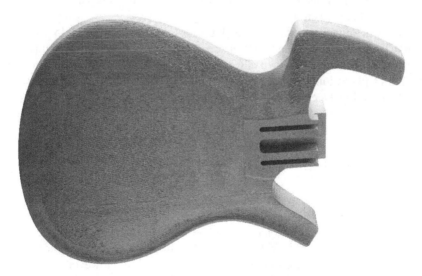

Fig. 3.43 Spruce Body Core for a Parker Fly Electric Guitar (Image courtesy of Parker Guitars, http://www.parkerguitars.com)

Fig. 3.44 Neck Intersection on a Parker Fly Electric Guitar (Image courtesy of Parker Guitars, http://www.parkerguitars.com)

among mass produced instruments. However, they have some advantages. The first is that they do not require a heel structure like that found on the Telecaster and Stratocaster. Thus, body contours can extend all the way to the neck, allowing easy access to the higher frets. Another is that the neck actually forms part of the body, so the instrument is very strong.

Alembic is known for producing neck through designs and Fig. 3.45 shows two Alembic basses made this way. The necks are made from three pieces of maple and two pieces of darker wood joined to form a single beam. The bodies are made from wings, thick plates joined to the sides of the neck block. These instruments clearly show the neck blocks extending all the way through the bodies.

3.4.1 Analyzing Necks

Simple intuition suggests that a guitar neck will change shape slightly under the tension of the strings. Most players prefer a slight upward curvature in the neck, usually called relief, and the natural deflection in response to string tension is sometimes counted on to provide it. This is particularly true on classical guitars that generally have no means of adjusting neck curvature. However, to develop a more rigorous understanding of the mechanics of guitars, it helps to have a means of at least approximately calculating the deflection of a guitar neck under load.

Fig. 3.45 Two Alembic Basses Showing Neck through Body Construction (Wikimedia Commons, image is in the public domain)

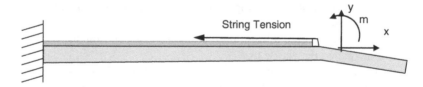

Fig. 3.46 An Idealized Guitar Neck

On almost all guitars, the neck is a discrete structure attached to the body and can be described mathematically using simple equations developed for beams. The analysis that appears here is greatly simplified and the results can be treated only as approximate. The point of including it is to show how even an approximate analysis can be helpful in understanding the structural behavior of guitars.

One end of the beam (the headstock) is treated as being free and the other end (the heel) is treated as if it is rigidly clamped to the body. This is called a fixed or cantilevered end condition [9]. Even on guitars with 'neck through' construction, where the neck is carved from a block that also runs the full length of the body, the neck can be considered as a beam cantilevered from the body. Figure 3.46 shows a diagram of a guitar neck with the string forces.

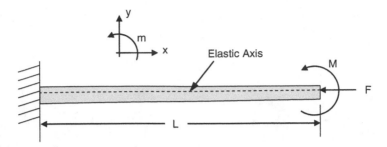

Fig. 3.47 Equivalent Guitar Neck Modeled as a Cantilevered Beam

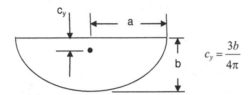

$$c_y = \frac{3b}{4\pi}$$

Fig. 3.48 Centroid of a Half Ellipse

In order to treat the neck as a simple cantilever beam, the string force acting above the neck is replaced by a force acting along the neck combined with a moment (the engineering name for a static torque) acting at the end of the beam. This new arrangement, shown in Fig. 3.47, is exactly equivalent, but easier to analyze. The compressive load, F, is simply the total tension of all six strings. The moment, M, is the total string force multiplied by the distance from the plane of the strings to a line called the elastic axis.

The elastic axis can be thought of as the line about which the beam deforms. Finding the exact location of the elastic axis requires an accurate geometric description of the cross section of the neck. Technically, the elastic axis is the line that passes through the centroids of cross sections of the neck.

The centroid is the point at which the cross section would balance if it were cut out of a uniformly thick plate. For our purposes, let's assume that the cross section of the neck is half an ellipse as shown in Fig. 3.48. Let's also assume that this is a classical neck with no truss rod as that simplifies the analysis.

The expression for the location of the centroid shows that it lies about ¼ ($3/4\pi =$ 0.239) of the distance from the top to the back of the neck. It is important to account for the extra stiffness from the fretboard. A simple way is to increase the effective depth of the neck as shown in Fig. 3.49.

It's also clear that the neck tapers along its length and this affects its stiffness. An accurate analysis requires accounting for the taper, but it adds significant complexity and isn't necessary here. Rather, we can assume the neck has a constant, representative cross section. For a numerical example, let's assume a representative

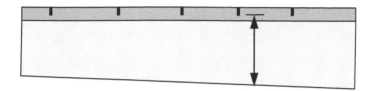

Fig. 3.49   Effective Thickness of the Neck with Fretboard

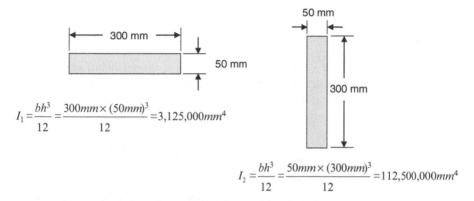

$$I_1 = \frac{bh^3}{12} = \frac{300mm \times (50mm)^3}{12} = 3,125,000mm^4$$

$$I_2 = \frac{bh^3}{12} = \frac{50mm \times (300mm)^3}{12} = 112,500,000mm^4$$

Fig. 3.50   Area Moment of Inertia of a Rectangular Beam

elliptical cross section 50 mm (1.97 in) wide and 18 mm (0.709 in) thick. The total tension for a typical set of classical guitar strings is 37.9 kg (83.6 lb or 372 N).

Compression of the neck due to the force, F, along the elastic axis is very small and is generally ignored. The expression for the deformation due to the moment acting at the end of the neck is

$$y(x) = \frac{Mx^2}{2EI} \tag{3.3}$$

where x is the distance along the beam. Mahogany has an elastic modulus, E, of approximately 9 GPa (1,305,000 psi).

The stiffness due to the cross-sectional shape is described by the area moment of inertia, I. The details of how to calculate the area moment of inertia for different shapes are an unneeded distraction here. It is enough to know that the higher the value of area moment of inertia, the stiffer the beam. A simple example is a floor joist, a beam that supports the floor in a house.

A typical floor joist might be 50 mm x 300 mm (approximately 2in x 12in). For a rectangular cross section, $I = bh^3/12$, where b is the width of rectangle, or its base, and h is the height of the rectangle. Figure 3.50 shows the effect on stiffness of installing the joist lying down or standing up.

Since $I_2/I_1 = 36$, turning the joist upright increases its stiffness by a factor of 36. This is why floor joists are installed standing up. If they were lying down, the floor

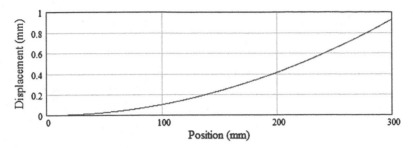

Fig. 3.51 Calculated Displacement of a Classical Guitar Neck

would be far too flexible – imagine walking down a diving board or across a trampoline. Note that material properties don't appear in the area moment of inertia; it is strictly a function of geometry.

The area moment of inertia of the half ellipse shown in Fig. 3.48 is

$$I = ab^3 \left(\frac{\pi}{8} - \frac{8}{9\pi} \right) = 0.1098ab^3 \tag{3.4}$$

This expression shows an important characteristic also present in the rectangular beam. The stiffness is proportional to the width of the beam, but a cubic function of height. This means that doubling the width of the beam doubles its stiffness, but doubling the depth of the beam increases its stiffness by a factor of eight. Put another way, small changes in the thickness of a guitar neck can significantly change the stiffness.

For the example of the 50 mm x 18 mm neck, I = 16,003 mm^4. Let's assume also that the combined tension of the strings is 372 N (83.6 lb) and the distance from the plane of the strings to the centroid is 8 mm. Figure 3.51 shows the calculated displacement.

The result is that the neck is predicted to deform approximately 0.9 mm at a point 300 mm from where it intersects the body. This is a small displacement compared to the dimensions of the instrument, but would be enough to create noticeable relief in the neck.

3.4.2 Truss Rods

Many guitars have an additional structural reinforcement in the neck in the form of a truss rod. Truss rods are either adjustable or non-adjustable. A non-adjustable, or fixed, truss rod simply adds stiffness to the neck while an adjustable one also allows the player to change the neck relief. Figure 3.52 shows a neck in cross-section with a fixed truss rod.

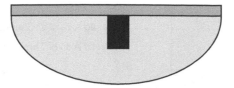

Fig. 3.52 Cross Section of a Neck with a Fixed Truss Rod

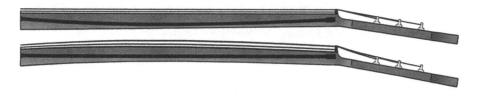

Fig. 3.53 Mechanics of a Single Acting Truss Rod (Wikimedia Commons, image is in the public domain)

Fixed truss rods are sometimes made from metal bars or rectangular tubes. Increasingly, though, they are made of unidirectional graphite. The elastic modulus of aluminum is about 10 times that of typical neck wood (such as mahogany) and that of steel is about 30 times that of wood. The elastic modulus of unidirectional graphite depends on several factors, but an average value is about 15 times that of wood.

The key idea is that a fixed truss rod can significantly increase the stiffness of a wood guitar neck. If the resulting neck is extremely stiff, it might be necessary to sand the desired relief into the fretboard since the neck will not deform noticeably under string tension.

Adjustable truss rods allow the curvature of the neck to be adjusted as desired. Adjustable truss rods are available in single acting and double acting designs. As the names suggest, single acting truss rods allow adjustment only in one direction and double acting one allows adjustment in both directions.

Single acting truss rods are common in steel stringed instruments and are generally installed so that they can only decrease curvature. This assumes that the strings will induce more curvature than desired and the difference will be removed by the truss rod.

The simplest type of single acting truss rod is essentially a long bolt mounted in the neck with an adjustable nut on one end. This design is often called a tension rod. The rod is almost always curved so that placing it in tension induces a crown in the center as shown in Fig. 3.53.

There are other types of single acting truss rods, but they all perform the same task. One problem with the simple tension rod shown above is that the pocket must be curved. It is easier to make a straight truss rod slot, so there is a need for a straight, single acting truss rod. One popular design, associated with Martin Guitars, uses a C shaped channel with a central tension rod as shown in Fig. 3.54.

Fig. 3.54 Cross Section of a C-Channel Truss Rod

![C-Channel Truss Rod under Tension]

Fig. 3.55 Cutaway View of C-Channel Truss Rod under Tension

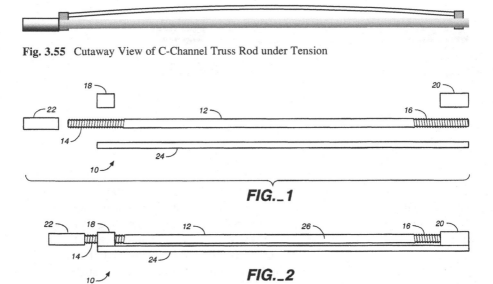

Fig. 3.56 Drawing of a Double Acting Truss Rod from Patent, 6,259,008

The mechanics of this type of truss rod are clever. The elastic axis of the C-channel doesn't line up with that of the threaded rod, so putting the rod in tension creates a moment that creates a crown in the C-channel. Fig. 3.55 shows a cutaway view with the rod under tension.

Finally, the double acting truss rod can either increase or decrease neck relief depending on which way the nut is turned. The most common form of double acting truss rod is shown in Fig. 3.56. These are two drawings from patent 6,259,008, Double-Action Truss Rod for Stringed Instruments, granted on July 10, 2001.

This design is elegant and simple, consisting of only five parts. The threaded shaft, part 12 in the drawing, has right hand threads (No. 16) on one end, and left hand threads (no. 14) on the other. The threaded blocks (parts 18 and 20) are brazed to a rectangular bar (part 24). Turning the nut (part 22) counterclockwise forces the blocks apart so that the assembly bows upward in the middle. Conversely, turning clockwise forces the blocks closer together so that the assembly curves downward. Like the C-channel design, this double acting rod can be installed in a flat bottom pocket.

3.5 Materials

There may be no part of guitar design more beholden to tradition than materials. By the mid 19th century, the conventional approach of a spruce top with rosewood sides and back was in place. Throughout most of the 20th century, acoustic guitars mostly followed this pattern, often using mahogany for necks.

Electric guitars were not in widespread use until after WWII so there was less influence from previous designs. Fender guitars were, from the beginning, made using maple necks with either alder or swamp ash for the bodies. The Gibson Les Paul was made using slightly more traditional materials. The body and neck were mahogany and the body often had a thick maple top plate.

The bottom line is the same for both modern acoustic and electric guitars; they are overwhelmingly made of wood. There are serious concerns about the continued availability of high grade wood for guitars, but efforts to establish sustainable production methods and a willingness to accept new species seem to be paying off. High grade wood is increasingly expensive, but still available in quantities sufficient to support global production.

3.5.1 Wood for Acoustic Guitars

The tops of acoustic guitars are mostly made from the wood of conifers, generally spruce, cedar or, less often, redwood. Sitka spruce and Englemann spruce are particularly popular, though other varieties of spruce are used. Spruce has a high ratio of stiffness to mass, making it particularly well-suited to the requirements of soundboards.

In an effort to reduce material costs, manufacturers began to use plywood for inexpensive acoustic guitars. While its use is still limited to inexpensive instruments that are often considered inferior to solid wood instruments, there doesn't seem to be any technical reason that laminated wood cannot be used to make top quality instruments.

Plywood has some very attractive properties. It is less susceptible to damage from changes in humidity. Solid wood tops sometimes split when subjected to very low humidity. Indeed, this author has had several very nice guitars suffer serious splits when displayed in his office. During the winter, the humidity there can be very low, sometimes less than 15%, and it is not possible to humidify only that space. Now, only solid body electrics and plywood acoustics are displayed and splitting problems have ceased.

Another advantage of plywood is that it can easily be formed while being laminated. Figure 3.57 shows a stack of laminated sides at the Taylor Guitar factory. They are in their final shape and did not need to be bent using heat. Also, since they have a cross grain ply in the center, they are very unlikely to ever split.

Aircraft grade plywood is also readily available. It is approved for load carrying structures on manned aircraft and is made to very high standards. It is very uniform and available in a range of thicknesses and materials. Like all plywood, it is

Fig. 3.57 A Stack of Laminated Guitar Sides (Image by the author, reproduced here courtesy of Taylor Guitars, http://www.taylorguitars.com)

Fig. 3.58 A Piece of Wood Being Book Matched

supplied in rectangular sheets that make it easy to machine and to incorporate in series production. The author has made a number of instruments using various types of aircraft grade plywood and had good results.

When solid wood tops and backs are used, the plates are usually book matched. This is a process in which a plank is split and opened up in the same manner as opening a book. Figure 3.58 shows a piece of zebra wood with a dramatic grain

Fig. 3.59 A Stack of Book Matched Back Plates (Image by the author, reproduced here courtesy of Taylor Guitars, http://www.taylorguitars.com)

pattern being book matched. The left image shows the board before being ripped into two pieces of equal thickness. The other two images show the two pieces after having been cut and then opened up for joining. The result will be a plate twice as wide as the original and with a grain pattern symmetric about the centerline.

Figure 3.59 shows a stack of book matched rosewood back plates at the Taylor factory. Notice the thin decorative strip inserted between the matched halves. Also, a close look also reveals a small triangular notch at the end of each plate next to the centerline. This is an alignment feature, included to ensure that the plates are precisely placed into production fixtures.

One convenient feature of plywood is that interior plies can be made of inexpensive wood that need only be structurally sound. Only the top ply will be seen, so it can be made of something particularly attractive. Since a single ply is much thinner than the completed plate, this is a way of making an attractive instrument while controlling material costs. Figure 3.60 shows the back of a Washburn WD-30 series acoustic guitar with a striking book matched, figured ash back. The plate is laminated so that only the visible ply needs to be of this select wood. This laminated back should be durable and stable, while still having the attractiveness of a figured and book matched plate.

From a structural standpoint, the most important characteristic of wood being used for necks and tops is the grain orientation. The most preferred orientation comes from quarter-sawn wood. The name derives from the way planks are sawn from logs. Figure 3.61 shows how a log can be split into quarters and then sawn into planks. The alternative is to slab saw the planks.

A good quarter sawn plank has straight grain lines perpendicular to the widest face. Such a board is much less likely to twist or otherwise deform with age. Figure 3.62 shows the grain of a nicely quarter sawn plank.

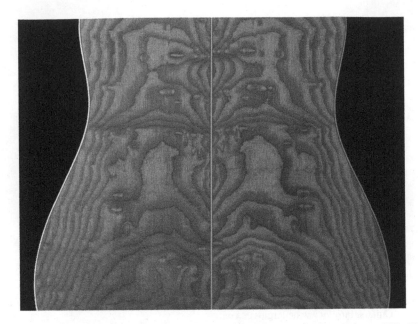

Fig. 3.60 A Laminated Guitar Back with a Book Matched Top Ply (Image courtesy of Washburn Guitars, http://www.washburnguitars.com)

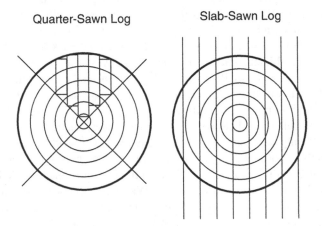

Fig. 3.61 Methods of Sawing Planks from Logs

3.5.2 Wood for Electric Guitars

The material requirements for solid body electric guitars are less stringent than those for acoustic guitars. The bodies, in particular, can be made from a range of

Fig. 3.62 Grain of a Quarter Sawn Plank

Fig. 3.63 A Washburn N4 with a Figured Koa Body (Image courtesy of Washburn Guitars, http://www.washburn.com)

relatively common woods. The typical solid body is 44.5 mm (1.75 in) thick and is much stronger than necessary to withstand string tension.

The wood chosen for solid bodies often depends on the finish. If a body is going to be painted with an opaque color, then grain and appearance aren't important and the wood can be selected for purely mechanical reasons. If it will be finished with a clear or translucent finish, then the appearance is also important. Sometimes, highly figured wood is used for guitar bodies in order to make a dramatic visual impression. Figure 3.63 shows a Washburn N4 made with a body made of highly figured koa.

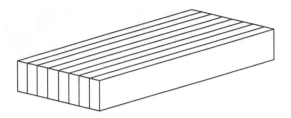

Fig. 3.64 An Ideal Quarter Sawn Board

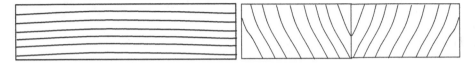

Fig. 3.65 Alternate Grain Orientations for Neck Blanks

Wood for electric guitar necks has the same requirements as that for acoustic guitars. The only major difference is that electric guitar necks tend to be longer and thinner, so the necks can be more susceptible to warping if inferior wood is used. Many different species can be used as long as the resulting planks are strong and straight, but maple and mahogany are probably most popular. This author has had very good experiences with beech and cherry as well.

The ideal neck blank is quarter sawn with the grain running parallel to the long axis of the board. Figure 3.64 shows the grain of an ideal quarter sawn board.

If quarter sawn lumber is not available, the next best thing is flat sawn lumber. Failing that, an acceptable neck blank can be made by joining two pieces of wood with the grain anti-symmetric about the glue joint. This way, internal forces that would otherwise cause the neck to warp are cancelled. Both of these are shown in Fig. 3.65.

3.6 Man Made Materials

Man-made materials have been used in guitars for some time, but are still not prevalent. Plastics are almost universally used for nuts, saddles and binding, but man-made materials are still not common in major components.

At first glance, it would seem that fiberglass or some other fiber reinforced plastic would be very suitable for acoustic guitars. It can be stored conveniently in rolls until needed. It is very uniform, so build variation due to material properties

Fig. 3.66 The Back of an Ovation Acoustic Guitar (Wikimedia Commons, image is in the public domain)

could be low. It can easily be molded into complex shapes. Finally, it is durable and resistant to humidity changes.

Acceptance of composite materials for acoustic guitars has been slow, in part due to changes in the tone of the resulting instruments. It is a common complaint that instrument with composite bodies sound different than do wood instruments. Some players object to this tonal change and some don't.

Perhaps the most well-known example is Ovation guitars and their use of composites for soundboards and for rounded, single piece backs. Figure 3.66 shows the back of an Ovation acoustic guitar. The back is made as a single piece in a shape far too complex to be formed from wood without great effort. Note the round access hole in the center of the lower bout. Removing the cover gives easy access to the interior of the instrument.

An interesting design using a composite back is the Yamaha APX-SPLI, shown in Fig. 3.67. It uses a plywood top with a nicely figured top ply combined with a molded back. The plastic body is lined with thin strips of a light wood that might be basswood. This is apparently to make the instrument sound more like a wood guitar. The result is a durable, attractive instrument with good tone.

A small number of manufacturers are making instruments exclusively from graphite. Probably the most successful of these to date is RainSong. Figure 3.68 shows a close up view of the upper bout of a RainSong JM-1000. The weave of the graphite cloth is clearly visible.

Fig. 3.67 A Yamaha APX-SPLI Acoustic Guitar with a Composite Back

Fig. 3.68 A RaingSong Graphite Guitar Showing Weave of Graphite Cloth (Wikimedia Commons, image is in the public domain)

References

1. Stratabond web site, http://www.rutply.com/solutions/stratabond.html, last visited 7/17/2011
2. ASTM Standard A228 (2007) Standard Specification for Steel Wire, Music Spring Quality, ASTM International
3. Bruné RE (2008) Classic Instrument: Antonio Torres 1863. Vintage Guitar
4. Kasha M (1969) Guitar Construction. Patent 3,443,465
5. Kasha M (1978) Bracing Structure for Stringed Musical Instrument. Patent 4,079,654
6. ASTM Standard D-905 (2004) Standard Test Method for Strength Properties of Adhesive Bonds in Shear by Compressive Loading, ASTM International
7. Johnston Rm Simmins M, Gerken T and Ford F (2005) Acoustic Guitar: The Composition, Construction and Evolution of One of the World's Most Beloved Instruments. Hal Leonard
8. Wadley L Hodgskiss T and Grant M (2009) Implications for Complex Cognition from the Hafting of Tools with Compound Adhesives in the Middle Stone Age, South Africa. Proceedings of the National Academy of Sciences.
9. Mott RL (2007) Applied Strength of Materials, 5th Ed. Prentice Hall

Chapter 4
Electronics

Guitar electronics can take many different forms. The first electronic components used with guitars were pickups and amplifiers. Since then, though, the development of electronics in other consumer products has been reflected in guitars. Indeed, features that were once only available in expensive electronic components are now being integrated into the on-board circuitry of electric guitars. These features include tuners, sound effects and recorders.

Fortunately, most components found in electric guitars are relatively simple and a basic understanding of how they work is enough to offer insight into the design and function of electric guitars. There are really only two classes of components found in most electric guitars: pickups and signal processing elements. A pickup simply converts string motion into a time-varying electrical signal (a time-varying voltage). The signal processing usually consists of volume and tone adjustments, but can include on-board effects, pre-amplifiers and even digital recorders. Note that 'signal' is just another name for a voltage that changes with time and has information encoded in it. The electrical signal coming from an electric guitar encodes the sound from the instrument. Not all time-varying voltages are signals; the 120 V AC power coming from wall outlet (220 V or 230 V in some places) is not usually referred to as a signal because it contains no information.

4.1 Electromagnetic Pickups

The first electromagnetic pickups were developed in the 1930s [1]. Perhaps the best known among the early pickups was the one used by Charlie Christian in the Gibson ES-150 (Fig. 4.1). It was very large by modern standards, being about six inches long and weighing almost two pounds. The ES-150 was first sold in 1936. Modern pickups are much smaller and often include features that reduce unwanted electromagnetic noise [2].

R.M. French, *Technology of the Guitar*, DOI 10.1007/978-1-4614-1921-1_4,
© Springer Science+Business Media New York 2012

Fig. 4.1 Gibson ES-150 with a Charlie Christian Pickup and the Pickup Itself (ES-150 image from Wikimedia Commons and is in the public domain. Pickup image courtesy of Seymour Duncan, http://www.seymourduncan.com)

Let's start by looking at the basic operation of a magnetic pickup. The principle that makes electromagnetic pickups work is Faraday's law of electromagnetic induction [3]. This law states that when an electromagnetic field moves with respect to a wire, then an electromotive force is induced in the wire. Electromotive force is the force that causes electrons to move through a wire and is measured in Volts, after Alessandro Volta who invented the battery.

We all are dependent on Faraday's law in our daily lives as it is the principle that makes electric motors and generators work. Figure 4.2 shows a bar magnet rotating next to a coil of wire. As the poles of the magnet pass the coil, they induce a voltage. The different poles will induce voltages of opposite signs, so the result will be a wave with both positive and negative peaks – alternating current or AC for short.

Using the same principle, applied in a different way, the magnetic pickup generates a time-varying voltage in response to a moving string. There is an elegant difference in pickups in that the coil is wrapped around a stationary magnet or a group of several stationary magnets. Without relative motion between the coil and the magnetic field, there is no signal, so this would seem to be a problem. However,

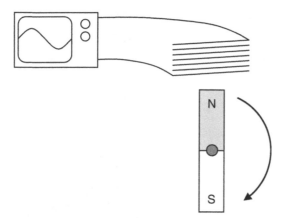

Fig. 4.2 A Simple Generator Making a Time-Varying Signal

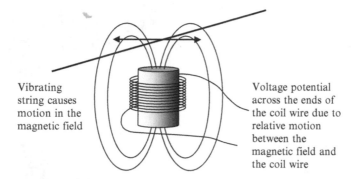

Vibrating string causes motion in the magnetic field

Voltage potential across the ends of the coil wire due to relative motion between the magnetic field and the coil wire

Fig. 4.3 Vibrating Wire Inducing a Voltage

the interaction between the magnet and the moving string is reciprocal; not only does the field affect the string, the string also affects the field, making it vibrate slightly. Since the field vibrates and the coil doesn't, a voltage is induced in the coil. That's the signal that is sent to the amplifier to make sound.

Figure 4.3 shows a single magnet and single coil in the presence of a vibrating string. However, the most common type of single coil guitar pickup uses six magnets with the poles all aligned in the same direction. The coil of wire is typically wound directly around the magnets or around a bobbin which holds the magnets.

Figure 4.4 shows the magnets for a traditional single coil pickup and the flat fiber plates (sometimes called flatwork) in which they are mounted. Very fine wire is wound around the magnets to form the coil that senses the vibrating magnetic field.

Figure 4.5 shows a typical completed pickup. The black and white wires are the ground and signal wires. The two ends of the coil wire can be seen emerging from the bobbin and connecting to the heavier black and white wires.

There are two common ways to mount a single coil pickup. One is to simply screw it to the bottom of the pickup pocket. The other method, used on the

Fig. 4.4 The Magnets and
Flatwork of a Single Coil
Pickup (Image courtesy
Stewart MacDonald, http://
www.stewmac.com)

Fig. 4.5 A Seymour Duncan
SSL-1 Single Coil Pickup
(Image courtesy Seymour
Duncan, http://www.
seymourduncan.com)

Stratocaster, is to hold the pickup in place by screwing it to the pick guard plate.
Figure 4.6 shows two single coil pickups in a Peavey Generation electric guitar. The
upper pickup is screwed to the guitar body and the lower pickup is screwed to the

Fig. 4.6 Two Single Coil
Pickups in a Solid Body
Guitar

bridge plate. Note also that the bridge pickup uses a three hole pattern for the mounting screws.

The Fender Stratocaster was designed to be easy to manufacture and this is apparent in how the pickups are mounted. All three pickups are mounted on the plastic pickguard along with the tone and volume knobs and a five way switch. Because the assembled pickguard forms a module, it can be installed on the body by simply screwing it on. Figure 4.7 shows two assembled pickguards, each with three pickups.

The variation in pickup designs and the large number of manufacturers means that there is no absolute standard for dimensions. However, tradition and the desire to make pickups that fit the most popular and iconic guitars means that there are some de facto standards. For example, single coil pickups are generally sized to fit Telecasters or Stratocasters. Double coil sized humbuckers are generally sized to fit Les Pauls. Figure 4.8 shows the dimensions, in inches, of the Seymour Duncan APS-2 single coil pickup.

The electromagnetic pickup is at once simple and elegant, but it does have one serious flaw; it is sensitive to electromagnetic noise. The world around us is bright with electromagnetic noise; cell phones, radio transmitters and electric equipment are everywhere and most of it emits electromagnetic radiation, some by design and some as a side effect. By far, the most important noise source for guitarists is the alternating current supplied through the power grid. This power gives off an electromagnetic signal with the same frequency as the power (60 Hz in the US and 50 Hz in most other places). A single coil pickup typically has something on the order of 6000 turns of wire and this coil acts as an antenna. An electrical signal

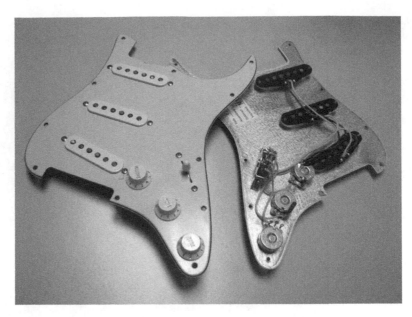

Fig. 4.7 Pre-Wired Pickguard Assembly for Stratocaster (Image courtesy Fender Musical Instrument Corp., http://www.fender.com)

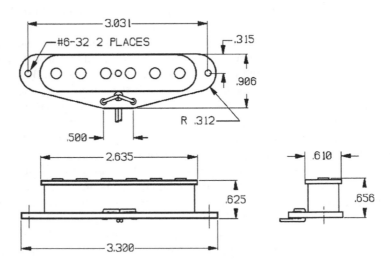

Fig. 4.8 Dimensions for a Typical Single Coil Pickup (Image courtesy of Seymour Duncan, http://www.seymourduncan.com)

generated by the pickup from electromagnetic noise is sent to the amplifier in the same way as the signal generated from string motion.

There is only one signal going to the amplifier and it has no way of distinguishing which parts of the signal came from string motion and which parts came from

Fig. 4.9 Coil Windings in a Humbucker

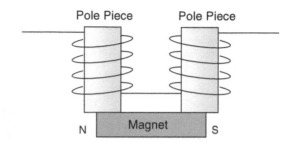

electromagnetic noise; it simply amplifies whatever signal is input. The 50 Hz or 60 Hz noise appears as a low frequency hum. In order to reduce hum, electric guitars have shielded electronics pockets and bridge grounds. Shielding prevents stray electromagnetic radiation from reaching the wires forming the circuit (though not the pickup coil). Grounding a circuit provides a path to zero electrical potential. It's approximately correct to think of ground as an infinite electron bucket; a circuit connected directly to ground is less noisy.

While shielding a circuit is a good way to avoid electromagnetic noise, another approach is to make a circuit that is not sensitive to electromagnetic noise to begin with. In 1955, Seth Lover, then working at Gibson, patented a design for a pickup that used two coils and was relatively insensitive to electromagnetic noise [4]. This type of pickup is universally known as a humbucker. Note that Rickenbacker had also developed a noise cancelling pickup and released it in 1953. However, it was not commercially successful and was withdrawn the following year.

The humbucker uses what we would now call common mode rejection between the two coils [5]. The coils are wound in opposite directions so that they are oppositely sensitive to a moving electromagnetic field. Thus, a moving field that would induce a positive voltage in one coil would induce a negative voltage in the other. The two coils are typically connected in series, so the noise signal in the first coil cancels that in the second. The only remaining problem is that the two coils would also be oppositely sensitive to string motion.

In order to make the two coils equally sensitive to string motion, the pole pieces for the two coils have opposite magnetic polarity – one north and one south. Since the magnetic fields have opposite signs, both coils are sensitive to string motion. The result is a pickup that is sensitive to string motion, but not electromagnetic noise. Figure 4.9 shows schematically how wire is wrapped around the poles on a humbucker.

There are now many variations on the basic humbucker concept, but the classic design is still quite popular. Figure 4.10 shows an exploded view of a humbucker. While this is a modern reproduction, it is essentially the same design introduced by Gibson in the 1950s.

Several things about this design are worth noting, as they became standard elements of most inductive pickup designs. We are conditioned to assume that the poles on a bar magnet are on the short sides of the magnet. On a humbucker, the magnet is polarized along the short axis, so that the poles are on the long sides.

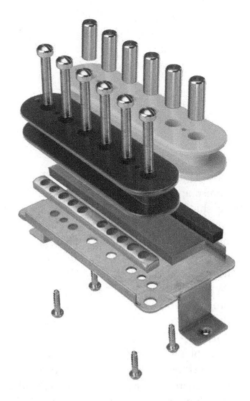

Since the magnet itself is below the two coils, small segments of steel called pole pieces are used to effectively extend the magnet to the top of the pickup. On a typical humbucker, one set of pole pieces is fixed while the other is threaded so that the individual pieces can be moved closer or farther from the strings.

Another important feature of the typical humbucker is how the wire coils are mounted in the pickup. Wire is not wrapped directly onto the magnets as with some single coil pickups. Rather, it is wrapped around plastic bobbins in which either magnets or pole pieces are mounted.

Another key design element is the tabs used to mount the pickup. Humbuckers usually have tabs on the sides of the base plate. The pickup is then mounted to a pickup ring using two screws with springs (or rubber tubing) around the screws in order to hold the pickup in place. The pickup ring is screwed, in turn, to the top of the guitar. Figure 4.11 shows the underside of a humbucker pickup mounted to a plastic pickup ring. Silicone rubber tubing was used here in place of metal springs. Figure 4.12 shows a Gibson Les Paul with two humbuckers mounted in rings.

The original commercial version of the Gibson humbucker was called the PAF (Patent Applied For) after a sticker that appeared on early models. By the time the patent was awarded, the name had stuck and they have been known as PAFs ever since. Currently, one of the most popular humbucker pickups is the Seymour Duncan '59, model SH-1, shown in Fig. 4.13. As the name suggests, this design is modeled closely on late 50's PAF humbuckers.

Fig. 4.11 The Bottom of a Humbucker Pickup with Mounting Ring

Fig. 4.12 A Gibson Les Paul with Two Humbuckers Mounted in Rings (Wikimedia Commons, image is in the public domain)

Fig. 4.13 Seymour Duncan '59 (Image courtesy of Seymour Duncan, http://www. seymourduncan.com)

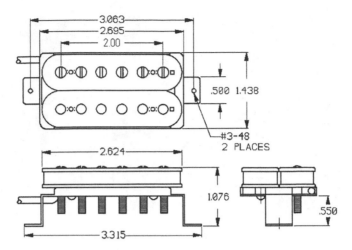

Fig. 4.14 Dimensions (in inches) of a Seymour Duncan '59 Humbucker (Image courtesy Seymour Duncan, http://www.seymourduncan.com)

As with single pole pickups, there is no official standard for the dimensions of humbuckers. Again, though, there is a defacto standard. Figure 4.14 shows the measurements of a Seymour Duncan '59 humbucker.

4.2 Pickup Circuit Models

It is quite helpful to have a simple electrical model of how a pickup works. However, we first need to understand the basics of voltage, current and impedance. Voltage was originally called electromotive force (EMF) and is a measure of the force pushing electrons through a wire. Current, expressed in Amps, is a measure of how many electrons flow through the wire each second. While it may initially seem odd, many people find it helpful to use analogies of water flowing through pipes when thinking of electrons flowing through wires. Voltage is analogous to the head pressure that causes water to flow through a pipe. Increasing the voltage in a circuit is equivalent to increasing the height of a water column, thus increasing pressure in a pipe. Current is analogous to how much water flows through a pipe. Increasing the diameter of a pipe allows more water to flow through it without having to increase pressure. Figure 4.15 shows the hydraulic analogies of EMF and current,

One other important concept in electronics design is impedance. Electrical impedance, designated by Z, is a measure of the resistance to the flow of alternating current. Impedance is the AC equivalent of resistance.

Impedance is an important parameter describing how electronic devices behave as a function of frequency. While DC resistance is usually constant, impedance generally varies with frequency.

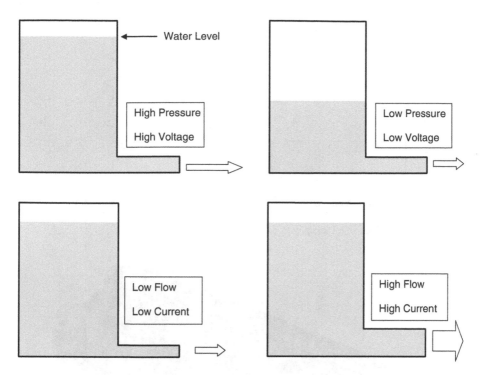

Fig. 4.15 Water Tanks Showing Hydraulic Analogies for Current and EMF

Impedance is also very important in calculating how components interact with one another. Just as it is difficult to generate a large force when pushing against something very soft, it is difficult to generate a large voltage drop across a component with low impedance – electrons need 'something to push against'. Even though impedance is an AC quantity, it is helpful to look at a DC analogy.

Figure 4.16 shows a very simple circuit with a voltage source and a resistor taking the place of the pickup and a resistor standing in for the load due to the amplifier. R_P is analogous to the output impedance of the pickup and R_L is analogous to the input impedance of the amplifier. Slightly rearranging the circuit shows that it is just a voltage divider circuit.

The output voltage of the voltage divider circuit is

$$V_{out} = \frac{R_L}{R_L + R_P} V_{in} \tag{4.1}$$

The amplifier can only work on the voltage it sees, so our goal is to make the output voltage as large as possible. In electrical terms, we want to drop as much voltage across the load as we can. The obvious way to do this is to make R_L large compared with R_P. Similarly, we want output impedance of a component to be small compared to the input impedance of the component it is feeding. A good guideline

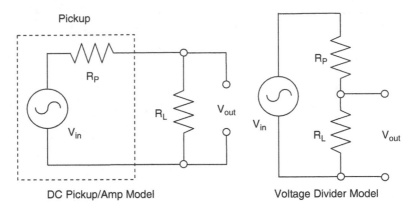

Fig. 4.16 DC Analogy of a Pickup with an Amplifier Load

Fig. 4.17 A Single Coil Sized Humbucker with Rails Rather than Individual Pole Pieces (Image courtesy of Seymour Duncan, http://www.seymourduncan.com)

is $Z_{in} \geq 10\,Z_{out}$. Thus, it is not surprising that the output impedance of pickups is generally in the range of 5 kΩ - 15 kΩ while the input impedance of amplifiers is generally in the range of 100 kΩ - 1MΩ. Let's start by looking at pickups.

In spite of the extremely simple design of inductive pickups, there are a number of design features that can be changed to affect the resulting sound. For example, the number of turns in the coils can be changed along with the diameter of the wire. Additionally, the strength of the magnetic field can be altered by changing the geometry of the magnet or the material from which it is made. The number and position of the pole pieces can be varied. Indeed, some pickups replace individual pole pieces with a single blade (often called a rail) spanning the width of the strings. Figure 4.17 shows a humbucker pickup with rails replacing the individual poles. While

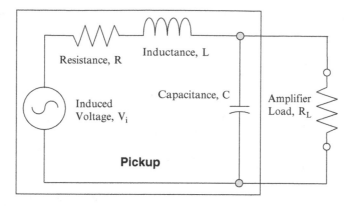

Fig. 4.18 Equivalent RLC Circuit for a Single Coil Pickup

this is a humbucker and has two coils so that it is not sensitive to electromagnetic noise, it is single coil sized and can be used in place of a single coil pickup.

The ideal pickup would exactly convert string motion into a proportional electric signal – a time varying voltage – and would be equally sensitive at all frequencies. However, all real pickups add their own signature to the motion of the string. In the strictest sense, they distort the ideal signal that would result from sensing the string motion. In popular terms, each pickup has its own tone.

If we are going to describe pickup tone and how it is affected by design choices, it is necessary to define some terms and to have a simple mathematical description of how a pickup works. The most important description of the pickup is its transfer function. The transfer function is the ratio of output to input, expressed in frequency domain.

$$H(\omega) = \frac{y(\omega)}{x(\omega)} \tag{4.2}$$

This is the standard notation for the transfer function. X is the input and Y is the output. We'll get to the mathematical description of X and Y shortly.

The key to making a mathematical description of the pickup is drawing an equivalent circuit. A complete electromagnetic model of a pickup and vibrating string is very complicated and is written in terms of differential equations. These are the equations that so famously strike fear into the hearts of engineering and physics students. Fortunately, it is possible to make a useful model of a pickup using idealized circuit elements and junior high school algebra.

A pickup by itself is not of much use; if the leads aren't connected to something, then there is an open circuit and no current will be able to flow. In order to close the circuit, an amplifier is connected to the pickup leads. The amp is also modeled using simple circuit elements and the result is called an equivalent circuit. The equivalent circuit diagram for a single coil pickup with amplifier is shown in Fig. 4.18. It is not exactly correct to model the amplifier as a simple resistance, since any real amplifier will have an equivalent input inductance and capacitance, but this approximation is sufficient for now. The additional mathematical complication is not needed.

The transfer function for this circuit is developed by starting with the Ohm's law [6] for alternating current.

$$V = iZ \tag{4.3}$$

Where i is current and Z is impedance. Recall that impedance is just the resistance to alternating current. Each of the components in the equivalent circuit has its own impedance. The impedances of inductors and capacitors are fundamentally different than for resistors in that they are frequency dependent and are not in phase with the current.

While the resulting expressions could be written in terms of trigonometric functions, they are cumbersome to work with; a much simpler approach is to use complex numbers to describe impedance. Complex numbers are just numbers with a real part and an imaginary part. In most mathematical writing, $i = \sqrt{-1}$, where i indicates an imaginary number. However, when describing circuits, i is reserved for current. In circuit analysis, the letter j is used for the complex variable, so $j = \sqrt{-1}$. The relationship between j and trig functions is given by the Euler equation

$$jx = \cos(x) + j\sin(x) \tag{4.4}$$

There is nothing at all intuitive about the Euler equation and it was a bit of a shocker in the mathematics world when Euler first developed it [7]. It is simply an identity like any other trigonometric identity. Its utility here is that it offers a way of using complex numbers – which are easy to manipulate – in place of trig functions – which aren't.

Imaginary may be the worst name ever for a class of numbers; it seems to imply that imaginary numbers are somehow not legitimate mathematical entities. They are just as legitimate as real numbers and very widely used in electrical engineering and many other technical fields. It would have been much better if real and imaginary numbers were, instead, called something less suggestive like red and blue numbers. If you want evidence that j is just a number, use a scientific calculator to show that $e^{j\pi} = -1$.

The expressions for impedance of the individual components are shown in Equation 4.5. Note that ω is frequency in radians/sec. 1 Hz $= 2\pi$ rad/sec and f $= \omega/2\pi$.

$$Z_R = R$$
$$Z_L = j\omega L$$
$$Z_C = \frac{-j}{\omega C} = \frac{1}{j\omega C} \tag{4.5}$$

Note that R is resistance measured in Ohms (Ω), L is inductance measured in Henries (H) and C is capacitance measured in Farads (F). With these expressions in hand, the task is now to use them to write an expression for the voltage in the equivalent circuit. If we re-group the elements slightly, the result is a circuit with a voltage source and two composite elements in series. A simple application of Ohm's law lets us write expressions for the voltage drops across the two composite elements. Impedances of the two elements are Z_1 and Z_2 as shown in Fig. 4.19.

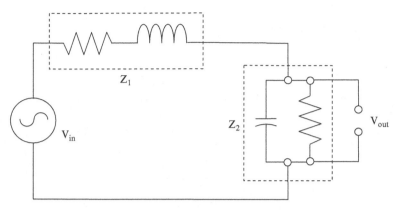

Fig. 4.19 Equivalent Circuit with Two Impedance Values

Using Ohm's law, $V_{in} = iZ_{Total}$ and $V_{out} = iZ_2$. Impedances in series add just like resistances in series, so $Z_{Total} = Z_1 + Z_2$. We can relate the input and output voltages by slightly re-arranging the expressions for the two voltage drops.

$$\frac{V_{in}}{Z_{Total}} = i \text{ and } \frac{V_{out}}{Z_2} = i \tag{4.6}$$

Since the circuit forms a closed loop, the current, i, is the same across both elements and we can set theses two expressions equal to one another.

$$\frac{V_{in}}{Z_{Total}} = \frac{V_{out}}{Z_2} \tag{4.7}$$

From there, it is simple to form the expression for the transfer function (the ratio of output to input).

$$H = \frac{V_{out}}{Z_{in}} = \frac{Z_2}{Z_{Total}} \tag{4.8}$$

Substituting in the expressions for the individual circuit elements gives the expression for the transfer function

$$H(\omega) = \frac{V_{out}}{V_{in}} = \frac{R_L}{(R + j\omega L)(j\omega C R_L + 1) + R_L} \tag{4.9}$$

Note that ω is the frequency in radians per second and that 2π radians/sec $= 1$ Hz. The transfer function written in terms Hz is

$$H(f) = \frac{V_{out}}{V_{in}} = \frac{R_L}{(R + j2\pi f L)(j2\pi f C R_L + 1) + R_L} \tag{4.10}$$

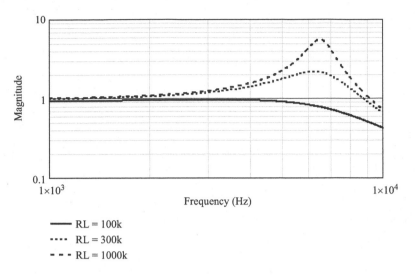

Fig. 4.20 Transfer Function of a Representative Single Coil Pickup

If the values of all the circuit elements are known, they can just be substituted into this expression. There are many single coil pickup designs and they generally have different electrical properties. A representative set of electrical properties are:

DC Resistance, R	7 kΩ
Inductance at 1 kHz, L	3 H
Capacitance, C	200 pf

Fig. 4.20 shows the transfer function calculated using Equation 4.10. The three curves correspond to three different input resistances for the amplifier. The two most significant features are the existence of a resonant peak and the effect of amplifier input resistance. The three characteristic parameters of the pickup, R, L and C, determine its resonant frequency. The amplitude and location of the resonant peak strongly affect the tone of the pickup. Much of the work of pickup design revolves around features that modify the pickup transfer function.

Our notional pickup has a resonant frequency of about 6200 Hz. The resonant peak has the effect of amplifying the components of sound near the resonant frequency. 6200 Hz is about five octaves above the frequency of the open G string and is a relatively high resonant peak for a guitar pickup. As a result, this pickup might be described as being bright – a reference to the increased high frequency content of the signal from the pickup.

Another important feature of Fig. 4.20 is the effect of changing the input resistance on the amplifier. Everything connected to the electric guitar necessarily becomes part of the circuit that produces the sound. When one device feeds a signal to another one, the concept of impedance matching becomes important. In electrical terms, we want the voltage drop across the amplifier to be as big as possible. Thus, we want the impedance of the amp to be large compared with the impedance of the circuit driving it.

As the resistance – the DC component of impedance – of the amplifier is reduced, the amplitude of the signal that reaches the amplifier is reduced. The amplifier circuit, even though it is modeled simply by a resistor, is coupled to the pickup and the transfer function of the circuit is dependent on both of them. The ideal amplifier should reproduce the signal coming from the guitar without changing it and the accepted way to do this is to make sure the input impedance of the amplifier is very high.

Ideally, the output impedance of the driving circuit should be about a factor of 10 lower than the input impedance of the circuit being driven. As the resistance of the amplifier gets very large, it approximates an open circuit; in practical terms, there is no need to increase input impedance beyond about 1 MΩ. There are far too many amplifiers available to draw overall conclusions, but a quick survey shows that specifications published for typical amplifiers list input impedances much higher than the DC resistance of any magnetic pickup.

Crate GFX50	270kΩ
Fender G-DEC Junior	>1 MΩ
Fender Pro Junior	> 1MΩ
Marshall JVM 210 H	470 kΩ

4.3 Pickup with Tone and Volume Controls

Pickups are almost never installed without a volume control and, usually, a tone control. The tone control is essentially a lowpass filter (low frequencies pass through it unchanged) that removes the high frequency, or treble, portion of the signal from the pickup. The type of filter used on guitars is sometimes called a treble bleed and is slightly different from the lowpass filter often shown in textbooks (Fig. 4.21).

It is convenient to think of this type of lowpass filter as a form of voltage divider. A traditional voltage divider has two resistors that work equally at all frequencies while the lowpass filter has a capacitor whose resistance is a function of frequency. At low frequencies, its resistance is very high, so V_{out} is high. At high frequencies, resistance is low, so V_{out} is low.

The treble bleed filter is different in that it has a resistor and capacitor in series with each other and in parallel to the voltage source. Because of this, it doesn't work when connected to an ideal voltage source as in Fig. 4.22. V_{out} is always equal to V_{in}.

In order for this tone circuit to work, it must be coupled to the pickup with its own impedance. Figure 4.23 shows a complete circuit including the pickup, tone and volume controls. Like other circuits shown here, it belongs to class of analog

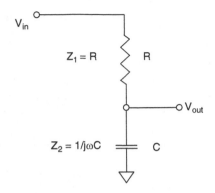

Fig. 4.21 Passive Lowpass Filter

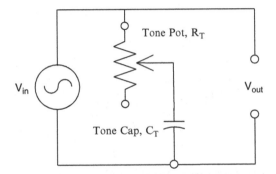

Fig. 4.22 A Treble Bleed Circuit in Parallel with an Ideal Voltage Source

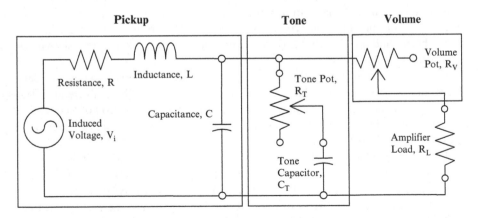

Fig. 4.23 Equivalent RLC Circuit for a Single Coil Pickup with Volume and Tone

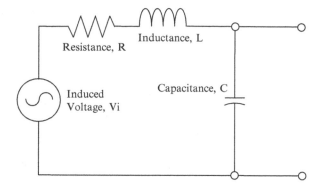

Fig. 4.24 Equivalent Circuit for a Pickup

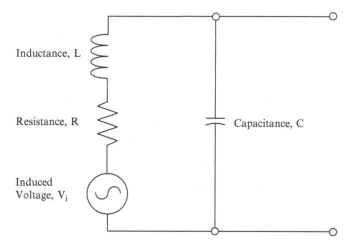

Fig. 4.25 Pickup Circuit Showing Parallel Connections

circuits made of resistors, inductors and capacitors. Their values are denoted by R, L and C respectively, so the resulting circuits are often called RLC circuits.

There are several ways to analyze this circuit. However, the most convenient is to rewrite it as a Thevenin equivalent circuit [6]. This approach replaces portions of the circuit with an equivalent voltage and equivalent impedance in series with each other. Let's start with the pickup itself as shown in Fig. 4.24.

Perhaps it isn't apparent that the voltage source, impedance and inductance are all in parallel with the capacitance. It is clearer when the circuit is slightly rearranged in a way that has no effect on its operation (Fig. 4.25).

The equivalent impedance, Z_{TH}, is the open circuit impedance with the voltage source replaced by a short. Z_P is just the parallel impedance of the Inductor and resistor on one leg and the capacitor on the other.

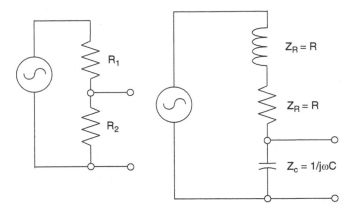

Fig. 4.26 The Pickup Circuit Drawn as a Voltage Divider

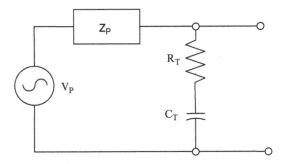

Fig. 4.27 Thevenin Equivalent Pickup Circuit with Tone Control Added

$$Z_P = \frac{(R + \omega L)\left[\frac{1}{j\omega C}\right]}{R + \omega L + \frac{1}{j\omega C}} = \frac{R + \omega L}{j\omega RC + j\omega^2 LC + 1} \qquad (4.11)$$

The equivalent voltage, V_P, is the voltage across the open terminals of the circuit. For this part of the calculation, it is convenient to think of the pickup as a type of voltage divider as shown in Fig. 4.26. The output voltage of a resistor voltage divider is $V_{out} = V_{in}R_2/(R_1 + R_2)$. Similarly, the output voltage for the pickup circuit is

$$V_P = \frac{\frac{1}{j\omega C}}{R + \omega L + \frac{1}{j\omega C}} V_{in} = \frac{1}{j\omega RC + j\omega^2 LC + 1} V_{in} \qquad (4.12)$$

The Thevenin equivalent circuit for the pickup shown in Fig. 4.27, now with the tone control added.

Fig. 4.28 Equivalent Circuit
with Volume and Amplifier
Load

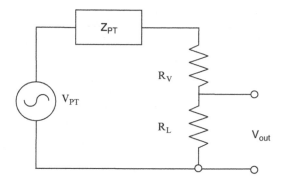

In order to account for the effect of the tone control on the circuit, let's Thevenize this circuit as well. As before, the impedances are in parallel, so

$$Z_{PT} = \frac{Z_P\left[R_T + \frac{1}{j\omega c_T}\right]}{Z_P + R_T + \frac{1}{j\omega C_T}} = \frac{Z_P(C_T R_T \omega - j)}{C_T R_T \omega + C_T Z_P \omega - j} \tag{4.13}$$

And, as before, the equivalent voltage is determined as if the circuit were a voltage divider

$$V_{PT} = \frac{R_T + \frac{1}{j\omega c_T}}{Z_P + R_T + \frac{1}{j\omega C_T}} V_P = \frac{C_T R_T \omega - j}{C_T R_T \omega + C_T Z_P \omega - j} V_P$$

$$= \frac{C_T R_T \omega - j}{(CLj\omega^2 + CRj\omega + 1)(C_T R_T \omega + C_T Z_P \omega - j)} V_{in} \tag{4.14}$$

Finally, the volume control and the load can be added to the new equivalent circuit on Fig. 4.28.

$$V_{out} = \frac{R_L}{Z_{PT} + R_V + R_L} V_{PT}$$

$$= \frac{R_L(C_T R_T \omega - j)}{(R_L + R_V + Z_{PT})(CLj\omega^2 + CRj\omega + 1)(C_T R_T \omega + C_T Z_P \omega - j)} V_{in} \tag{4.15}$$

The goal of this algebraic exercise is to find the transfer function, H, where $H = V_{out}/V_{in}$. It's clear from the previous expression that

$$H = \frac{V_{out}}{V_{in}}$$

$$= \frac{R_L(C_T R_T \omega - j)}{(R_L + R_V + Z_{PT})(CLj\omega^2 + CRj\omega + 1)(C_T R_T \omega + C_T Z_P \omega - j)} \tag{4.16}$$

Table 4.1 Pickup Circuit Parameters

Description	Symbol	Value
Pickup Resistance	R	7000 Ω
Pickup Inductance	L	3 H
Pickup Capacitance	C	200 pF
Tone Capacitance	C_T	0.022 μF
Tone Resistance	R_T	0-500 kΩ
Volume Resistance	R_V	0-500 kΩ
Load Resistance	R_L	500 kΩ

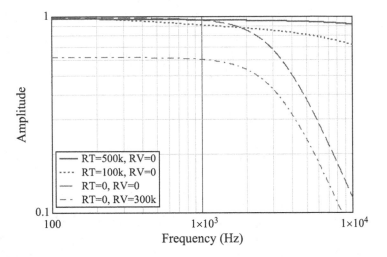

Fig. 4.29 Transfer Function for Pickup with Tone and Volume Controls

This expression might make more sense if we use some representative numbers and plot the transfer function through the whole circuit. Let's use the numbers from the previous plot along with some representative numbers for tone and volume controls. The circuit parameters are presented in Table 4.1 and calculated transfer functions are shown in Fig. 4.29.

4.4   Pickup Tone

A continuing source of confusion with pickups is the language used to describe tonal quality. People naturally resort to subjective, vague terms to describe sound and it can be difficult to translate this 'dude speak' into objective terms.

The first distinction to define is high output vs. low output. The voltage induced in a coil is proportional to the number of turns in the coil and to the number of magnetic field lines cut per second. As the field vibrates, lines of constant magnetic flux pass through the coil. Figure 4.30 shows a 2-D slice of the magnetic field calculated around a single pole pickup with two magnets – akin to a Gibson P-90.

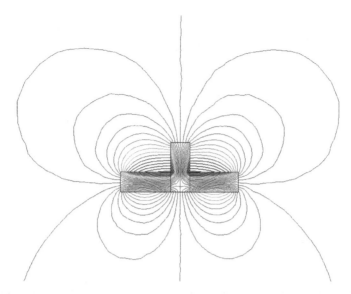

Fig. 4.30 2-D Magnetic Field Calculated for a Single Pole Pickup with Two Magnets

A high output pickup is one with more turns of wire (usually of a smaller diameter), stronger magnets or both, but this comes at a price. As wire diameter decreases and the length of the wire increases, impedance and capacitance increase. Also, more turns of wire make the pickup more susceptible to electromagnetic noise. An alternative method for increasing output is to include a small pre-amp to boost the signal from one or more coils with relatively few turns; this allows a higher signal level with reduced sensitivity to electromagnetic noise.

One of the most important characteristics of a pickup is the location of the resonance peak [8]. Pickups described as 'bright' usually have resonant peaks in the 4 kHz – 8 kHz range. Raising the resonant peak further can have unintended effects; if the resonant peak is about 10 kHz, the pickup can actually sound less bright because hearing sensitivity tends to drop at higher frequencies (peak sensitivity for humans is generally near 1000 Hz). Figure 4.31 shows an A-weight curve intended to approximate the sensitivity of humans to sounds of different frequencies [9].

Pickups with resonant frequencies below 4 kHz can have a pronounced midrange character. If the pickup has strong output, the result can be a distinctive sound sometimes described as 'tanky' (taken from a resonant tank). Conversely, lowering the resonance frequency and reducing the power of the magnet can result in a 'warm' sound. Reducing magnetic strength has the effect of reducing the initial attack – the rate at which sound builds after the string is plucked.

Another element over which the designer has control is the choice of magnet materials. The range of choices is theoretically quite broad, but there are only a few materials in widespread use. The most popular are Alnico 2, Alnico 5, Alnico 8 and

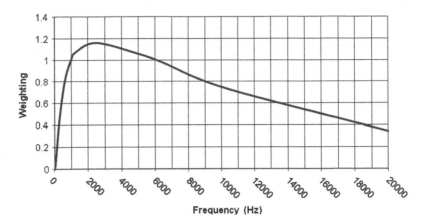

Fig. 4.31 An A-Weighting Curve

ceramic composition [10]. High magnetic strength materials like Neodymium and Samarium Cobalt are used in small, powerful electric motors. However, they are still rare in guitar pickups.

While the basic magnetic pickup is made of just wire, magnets and a few small pieces of metal, there is a wide range of design choices that can affect how pickups sound. The designer can choose from many different geometries and different materials in order to create different electrical characteristics and, thus, different sounds. Figure 4.32 shows a collection of different guitar and bass pickups that suggests some of the range of designs currently available.

Let's start with geometry. The basic idea of an inductive pickup is that a magnetic field vibrates in the presence of a coil of wire. The magnets in the pickup make a static field since the magnets don't move. The field moves only when the string vibrates. Since the coil isn't moving, there is relative motion between the vibrating field and the static coils, and a voltage is generated.

There are several common coil and magnet geometries. Of these, the oldest and the simplest is the single coil pickup. Figure 4.33 shows a classic single coil pickup as introduced on the Fender Telecaster and Fender Stratocaster. Six magnets are held in an evenly-spaced pattern by two pieces of flat fiberboard (colloquially called flatwork) and a single coil is wound around them using fine, insulated copper wire. The two ends of the coil wire are then connected to heavier black and white wires that, in turn, go to the output jack.

In even this simplest of pickups, there are many design choices to be made: the six magnets can be made of several different materials; the dimensions of the magnets can be changed; the number of turns in the coil can be changed; and even the size of the coil wire can be changed. All of these seemingly minor changes can noticeably affect the sound produced by the pickup.

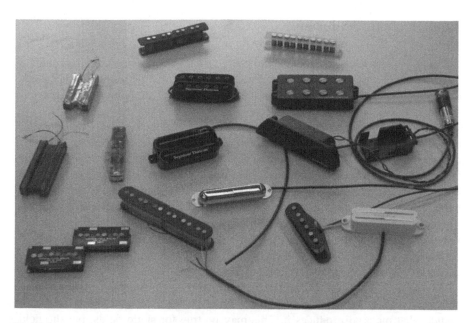

Fig. 4.32 A Collection of Electromagnetic Pickups Showing a Variety of Different Designs (Image courtesy of Kevin Beller, Seymour Duncan, http://www.seymourduncan.com)

Fig. 4.33 A Classic Single Coil Pickup (Image courtesy of Kevin Beller, Seymour Duncan, http://www.seymourduncan.com)

Let's take these possible design variations one at a time, starting with the magnet material. Materials capable of generating a strong magnetic field are surprisingly recent. In fact, one reason that the magnets on the Charlie Christian pickup in Fig. 4.1 are so large is that materials available at the time didn't produce a strong magnetic field. Loosely speaking, we can divide modern magnetic materials into three categories: ceramic, alnico and rare earth.

4.5   Magnetic Materials

Ceramic magnets are also called ferrite magnets and generally contain iron oxide –
either hematite (Fe_2O_3) or magnetite (Fe_3O_4) - and barium or strontium oxide. They
don't produce a terribly strong field, but are inexpensive and maintain their magnetism over a long period of time.

Alnico is a metal composed of aluminum, nickel and cobalt with smaller
amounts of copper and sometimes titanium. There are different versions of alnico
designated by a number, such as alnico 5 and alnico 8. For example, a pickup may
be listed as having alnico 5 magnets. There is no simple, accurate way to relate the
different grade numbers to physical properties. Suffice it here to say that
the different grades result in pickups with different acoustic properties.

Rare earth magnets typically contain samarium and neodymium. They generate
very strong magnetic fields and are not now widely used in guitar pickups. Figure 4.34
shows two small neodymium magnets placed on the author's finger. The bottom
magnet is suspended only by attraction to the one on the top.

At first glance, it might seem that the material used for the magnets in a pickup
shouldn't matter. After all, a magnetic field is a magnetic field and it shouldn't
matter what material produces it. This may be true for static fields, but the field
surrounding a pickup is dynamic – it moves in response to the motion of the strings.
Different magnetic materials have different dynamic behaviors [11] and it follows
that different magnet materials create different characteristic tonal qualities. General guidelines [12] are:

Alnico

- Available in grades 1-9, though 2, 5 and 8 are most widely used for pickups
- Ferrous content increases inductance compared to ceramic magnet core or air
  core. Increased inductance lowers resonant frequency.

Fig. 4.34 Demonstration of the Strength of Neodymium Magnets

- Alnico 2 is isotropic and can be magnetized in any direction. Field is lower strength and diffuse. Provides softer attack and lower output.
- Alnico 5 and 8 are anisotropic and have to be magnetized along a preferred direction. Field strength is higher than Alnico 2. The output is higher and the attack is more pronounced.

Ceramic

- Has little effect on coil inductance. Low ferrous content makes it behave, for practical purposes, like an air core.
- In theory, ceramic magnets have lower field strength than alnico, assuming ideal geometries. However, design constraints mean that ceramic magnets in guitar pickups can have higher field strength than alnico 2 and 5.

Neodymium

- Very high field strength
- Sensitive to heat
- Still not commonly used in pickups, but can be found in microphones and speakers

Samarium Cobalt

- Extremely high field strength
- Better heat resistance
- Rarely used in pickups, but can be found in high performance motors

4.6 Pickup Geometry

The most obvious design variation is the geometry of the coils and the magnets. The voltage induced by the string motion is proportional to the number of turns in the coil and the rate at which the coil cuts lines of constant field strength [13]. The mathematical expression is

$$V = N\frac{d\phi}{dt} \tag{4.17}$$

Where N is the number of turns and $d\phi/dt$ is the rate at which the coil cuts the field lines (which are moving).

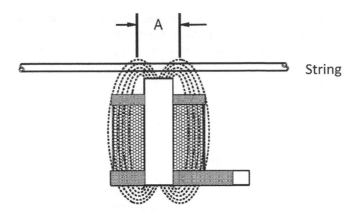

Fig. 4.35 Cutaway of a Single Coil Pickup Showing Field Lines (Image courtesy of Kevin Beller, Seymour Duncan, http://www.seymourduncan.com)

To see how this all works, it is helpful to look at a diagram showing a cutaway coil with magnets and field lines superimposed. Figure 4.35 shows a cutaway of a single coil pickup ('Strat Style') like that in Fig. 4.33. The cutaway shows a single magnet with the magnetic axis oriented vertically. The dotted lines show lines of constant magnetic field strength.

The length, A, is called the window length. This is the portion of the string observed by the pickup. The pickup essentially averages the motion of the string over this length, and averaging has the effect of attenuating high frequencies. It is analogous to applying a low-pass filter. Roughly speaking, the larger the value of A, the more the motion of the string is averaged and the more high frequencies are removed. Conversely, the shorter the window length is, the less the motion is averaged and the more high frequency content is preserved. In practice, this means that single coil pickups – or at least single coil sized pickups – are often described as sounding brighter than humbuckers, which are larger.

Fig. 4.36 shows a cross section of a typical humbucker. The magnet is placed between two sets of poles, one fixed and one made with screws to be adjustable. The humbucker is much less susceptible to noise, but has a much larger window length. Not surprisingly, they are not generally considered as bright as single coil pickups.

One response to this problem is to merge the two design concepts. There are a number of pickups now available that have two coils and a shared magnet like the standard humbucker shown above, but have the dimensions of a single coil pickup. Figure 4.37 shows the top and bottom of a Seymour Duncan Hot Rail pickup. It is the same size as a typical single coil pickup while still being hum-cancelling. The top picture clearly shows the two coil bobbins and the bottom picture shows the magnet between the two rails. Remember that the magnet is polarized along the short axis.

One obvious feature of this design is that the individual pole pieces have been replaced by plates, usually called rails or blades. These rails set up a magnetic field that has relatively little variation from side to side. This means that the player can bend

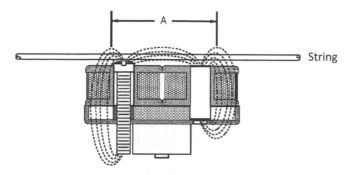

Fig. 4.36 Cutaway of a Humbucker Pickup Showing Field Lines (Image courtesy of Kevin Beller, Seymour Duncan, http://www.seymourduncan.com)

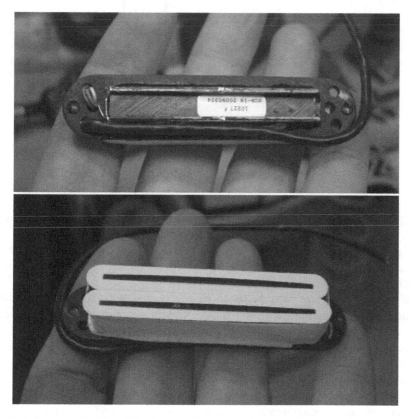

Fig. 4.37 A Single Coil Sized Humbucking Pickup with Rail Pole Pieces

Fig. 4.38 Cross-section of a
Single Coil Sized
Humbucking Rail Pickup
(Image courtesy of Kevin
Beller, Seymour Duncan,
http://www.seymourduncan.
com)

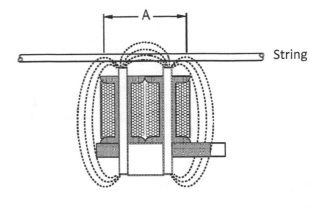

Fig. 4.39 Cross Section of a
Modified Vintage Rail Pickup
(Image courtesy of Kevin
Beller, Seymour Duncan,
http://www.seymourduncan.
com)

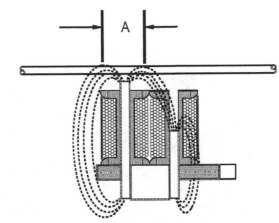

strings (pull them parallel to the fret in order to increase tension and, thus, pitch) with less worry about moving the string out of the field of a pole piece.

Figure 4.38 shows a cutaway of a Seymour Duncan Hot Rail pickup. Because of the smaller width, the window length is also smaller and more of the high frequency content is preserved.

Of course, there are many different magnet and coil configurations possible and designers have latitude to change their designs to accommodate specific needs. One goal is to further reduce the window length while cancelling hum. In response, the Seymour Duncan Vintage Rail pickup has rails of unequal heights as shown in Fig. 4.39. The string interacts strongly with the left hand rail and weakly with the right hand rail, so the window length is nearly the same as for a single coil pickup. However, the presence of two rails and two coils means that there is still a hum-cancelling effect.

There are far too many geometries to describe here, but one last arrangement to consider is the stacked pickup as shown in Fig. 4.40. In this type of design, one coil is placed above the other and both surround a single set of pole pieces. This design is intended to offer single coil sound while still cancelling electromagnetic noise (hum).

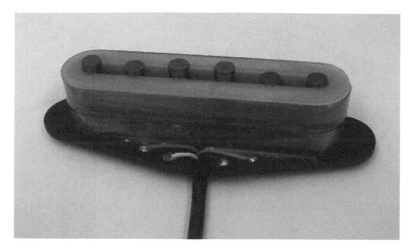

Fig. 4.40 A Seymour Duncan Classic Stack with Cover Removed to Show Coils

4.7   Other Advanced Magnetic Pickup Designs

Until now, the focus has been on basic electromagnetic pickups with simple magnet geometries and large numbers of turns of thin wire wrapped around a bobbin or even the magnets themselves. The basic designs for single coil and humbucking pickups were well established by 1960. Since then, however, many alternative designs have been proposed and some have been commercially successful.

The ideal pickup would accurately sense the motion of the string without affecting how the string behaves. It would also be very sensitive so it would still produce a useful signal during quiet passages when the amplitude of the string motion was low. Finally, it would be equally sensitive at all frequencies. No real pickup can do all these things, so designing pickups always requires compromises between opposing requirements.

The sensitivity of a passive pickup is a function of the strength of the magnetic field and the number of turns in the wire coils. The sensitivity from the pickup is directly proportional to the number of turns in the coil(s), so doubling the number of turns doubles the sensitivity of the pickup. Increasing the strength of the magnetic field increases the inductance, L, of the pickup and, thus, output signal.

The total effect of increasing the strength of the magnetic field or increasing the number of turns in a coil is more complex than it may first appear since the pickup and string form a system of interacting elements. For example, increasing the strength of the magnetic field will increase the level of the signal induced in the coil, but a strong field can also affect string vibration. Remember that energy is conserved. The energy created in a circuit has to come from somewhere; in this case, it comes from the kinetic energy in the string.

Fig. 4.41 Seymour Duncan AHB-1 Blackout Active Pickup (Image courtesy of Seymour Duncan, http://www.seymourduncan.com)

Similarly, increasing the number of turns necessarily creates some side effects. The increased length of wire increases the resistance and the increased number of turns nestled next to each other increases the capacitance. Changes in any of the parameters that comprise the pickup circuit can change the frequency response of the pickup. To the player, this means changes in the tone of the pickup.

One approach to improving pickup performance is the active pickup. Traditional pickups consisting only of magnets, wire, metal pole pieces and plastic bobbins are sometimes called passive pickups. They don't need external power and have no internal circuitry other than the wire that forms the coil(s). Conversely, an active pickup typically has a small solid state amplifier (a pre-amp) built in and might even have other circuit elements that modify the output signal. Figure 4.41 shows two Seymour Duncan AHB-1 Blackout active pickups.

The addition of a pre-amp to a pickup offers several design options that aren't practical with a passive pickup. For example, it is possible to decrease the number of turns in the coil to make it less sensitive to electromagnetic noise. The output amplitude is proportionately decreased, but this can be corrected by the pre-amp. Active pickups are often described as being quieter and cleaner than passive pickups. This is due in part to the freedom the designer has to change the number of coils and the diameter of the wire used. Another benefit of the pre-amp is that it allows the designer to change the output impedance of the pickup independently of the coil characteristics. Finally, on-board filtering circuits can be used to modify the tone electronically and make a single design very versatile.

There are also advanced passive pickup designs. These generally use the same types of parts seen in more conventional designs, but incorporate specially designed components to shape the magnetic field. An early example is the Lace Sensor, developed by Don Lace and originally produced in 1985 [14,15]. This pickup is similar to other single coil designs, but has a shield fitted around the coil to both isolate it from electromagnetic noise and to shape the field lines (Fig. 4.42). The result is a single coil pickup with a clear sound and greatly reduced hum.

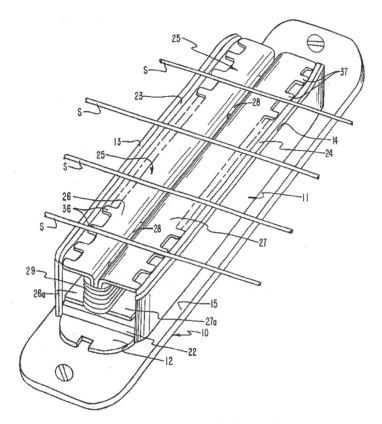

Fig. 4.42 Drawing from Patent 4,809,578 (Filed in 1987 by Donald Lace) Showing an Early Single Coil Design with Shielding and Notched Pole Plates

More recent designs combine thin plates to tailor the electromagnetic field with secondary noise cancelling coils. An example is the Seymour Duncan Stack Plus, a passive pickup designed to offer the tonal quality and packaging of a single coil pickup while still being noise cancelling. Figure 4.43 shows a cutaway of this design.

Figure 4.44 shows an exploded view of the complete pickup, except for the coils, which are omitted for clarity.

Electrical transformers, which increase or attenuate the amplitude of dynamic signals using electromagnetic induction, are almost always made with stacks of thin ferrous plates. This is to limit efficiency losses due to eddy currents. In the same manner, it is possible to make electromagnetic shielding elements in a pickup using stacked plates. Figure 4.45 shows a concept for a combination lower coil form and shield made this way.

Figure 4.46 shows an exploded view from patent 7,227,076 of a pickup designed for rare earth magnets. This is configuration is used in the Fender SCN, one of the few commercially available pickups to use rare earth magnets, samarium cobalt in

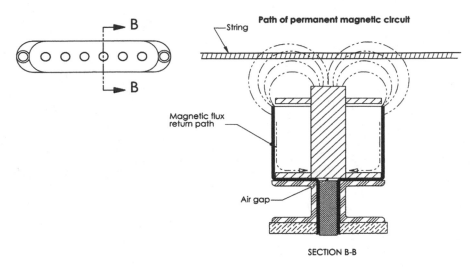

Fig. 4.43 Cross Section of a Stacked Pickup with Flux Transfer Plate Showing Magnetic Field Lines (Image courtesy of Kevin Beller, Seymour Duncan, http://www.seymourduncan.com)

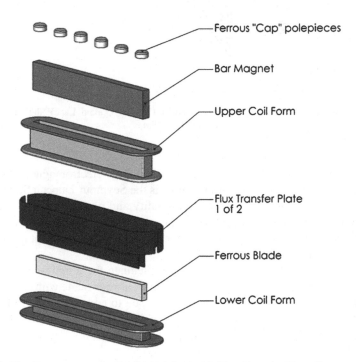

Fig. 4.44 The Components of a Stacked Pickup with Flux Transfer Plate (Image courtesy of Kevin Beller, Seymour Duncan, http://www.seymourduncan.com)

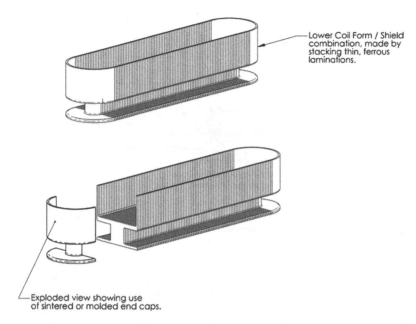

Lower Coil Form / Shield combination, made by stacking thin, ferrous laminations.

Exploded view showing use of sintered or molded end caps.

Fig. 4.45 A Combination Lower Coil Form and Shield Made of Stacked Plates (Image courtesy of Kevin Beller, Seymour Duncan, http://www.seymourduncan.com)

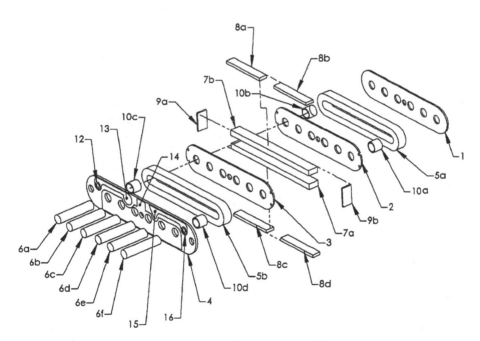

Fig. 4.46 Patent Drawing Showing Components of a Pickup Designed to Use Rare Earth Magnets

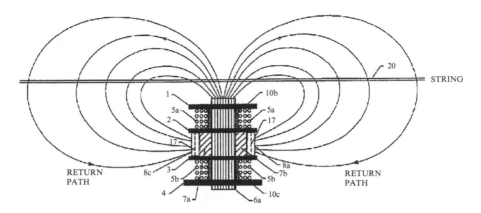

Fig. 4.47 Patent Drawing Showing Components the Magnetic Field

this case (elements 8a through 8d). This design uses non-magnetized, ferrous pole pieces (parts 6a through 6f) as seen in many traditional pickups. There are also two coils (5a and 5b) so that the pickup can reject noise.

Perhaps the most interesting feature is the placement of the magnets. They are so strong, they don't need to be very big. They are placed on ferrous moderator bars (7a and 7b) that serve to distribute the magnetic field. Figure 4.47 shows the magnetic field from the magnets, moderator bars and pole pieces.

4.8   Piezoelectric and Seismic Pickups

The first widely used pickups were electromagnetic, using the vibrating strings to induce a signal in a coil of wire placed in a magnetic field. Indeed, these are still the most common. However, there are other ways of sensing string vibrations and making a proportional electrical signal.

Electromagnetic pickups are universally used on solid body electric guitars, but seldom on acoustic guitars. Being made of magnets and wire, electromagnetic pickups are, by definition, heavy. Mounting one directly to the top of an acoustic guitar would significantly increase the mass of the top, making it much harder for the strings to vibrate it. The result is likely to be a guitar with poor sound when not amplified.

An alternative approach for acoustic guitars is to use a pickup that senses motion of the top or something related to it rather than motion of the strings. There are basically two common ways of doing this. The first is to sense the motion of the top directly and the second is to sense the force between the strings and the bridge.

One way to sense the motion of the top is to use an accelerometer – a device that turns acceleration into a proportional voltage [16]. A typical accelerometer uses a seismic mass attached to a mounting pad through a force sensor as shown in

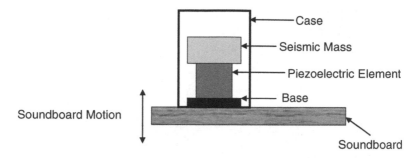

Fig. 4.48 Parts of a Piezoelectric Seismic Accelerometer

Fig. 4.48. The seismic mass is typically just a carefully machined piece of metal. The force sensing element is usually piezoelectric – either a ceramic element or a thin polymer film.

A piezoelectric material generates a voltage potential when it is deformed. The piezoelectric effect was first demonstrated in natural crystals in 1880 by the brothers Pierre and Jacques Curie [17]. Their materials included tourmaline, topaz, quartz and cane sugar. Quartz was shown to strongly exhibit the piezoelectric effect and found one of its first practical applications in underwater sonar transducers. Natural quartz is not a convenient engineering material so piezoelectric ceramics and plastics were developed. Perhaps the most common piezoelectric ceramic is lead zirconate titanate, usually abbreviated as PZT [18]. The only piezoelectric polymer in wide use is polyvinylidene fluoride, usually abbreviated as PVDF.

If the base of the accelerometer accelerates, the piezoelectric element is deformed. If the case accelerates upward, the piezoelectric element is compressed between the mass and the base – remember that $f = ma$ and the mass will accelerate along with the case only if the piezoelectric element pushes it. If the case accelerates downward, the piezoelectric element will be pulled between the base and the mass. In either situation, the piezo element makes a charge proportional to the acceleration.

The proportional charge made by the piezo element must be converted to a proportional voltage in order to provide an input signal to the amplifier. Commercially available accelerometers designed for dynamic structural testing typically have a small amplifier built into the base and require a power supply.

A simple piezo element with connecting wires acts as a voltage source with a very high output impedance – on the order of 1 MΩ. If connected to a guitar amplifier designed to work with pickups having 5kΩ – 15kΩ output impedance, the amplitude of the signal will be greatly reduced. The usual solution is to place a pre-amplifier between the piezo sensor and the amplifier. In addition to gaining up the signal from the sensor, the pre-amp will solve the impedance mismatch problem. A well-designed preamp has very high input impedance, so it can be driven by a piezo element, and very low output impedance, so it can, in turn, drive a guitar amp.

Another way to amplify an acoustic guitar with a piezoelectric element is to fix it directly to the soundboard. The slight deformation of the soundboard deforms the

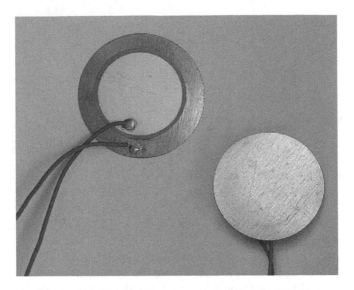

Fig. 4.49 Piezo Discs (Wikimedia Commons, image is in the public domain)

sensor, making the output signal. A common type is the piezo disk. Piezo disks are very inexpensive, light and compact, so they are easy to package inside a guitar. Figure 4.49 shows two piezo discs. The back plane serves as the ground and the lighter, inner circle is the voltage source.

A commercially available piezoelectric soundboard pickup is the DPU-3 by Seymour Duncan as shown in Fig. 4.50. It consists of a piezo film accelerometer and a compact pre-amp that is easily mounted inside the instrument.

By far, the piezo sensors most commonly used in guitars are those mounted under the saddle. These are either thin bars of piezoelectric material or braided flexible cable with a piezoelectric component. Figure 4.51 shows a flexible cable pickup with a 2.5 mm plug to connect with a pre-amp.

The cable is mounted under the saddle so that it senses the pressure between the saddle and the bridge. There is a large static force acting on the saddle and pickup because of the string tension. However, there is also a small dynamic force due to the string vibration. Figure 4.52 shows a bridge with the pickup inserted through a hole drilled at the end of the saddle slot and a cross section showing the sensor between the saddle and the bridge.

The output impedance of piezo cable pickups is generally high enough that a pre-amp is required to lower it before signal can go to the amplifier. However, with careful design, it is possible for a piezo sensor to have low enough impedance that it can drive a guitar amplifier. Figure 4.53 shows a Sadducer™, a piezoelectric under saddle sensor that does not require a pre-amp.

There is a subtle, but important, consideration in choosing pickups for acoustic guitars. Acoustic guitars make sound when the soundboard moves and creates pressure waves in the surrounding air. More specifically, the pressure waves are produced

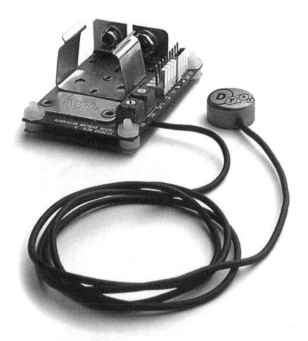

Fig. 4.50 A Soundboard Pickup with Pre-Amp (Image courtesy of Duncan-Turner Acoustic Research, http://www.d-tar.com)

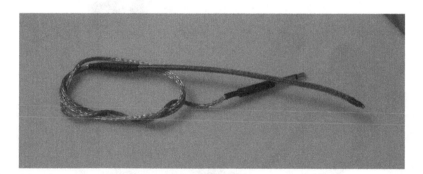

Fig. 4.51 Flexible Piezo Cable Under-Saddle Pickup

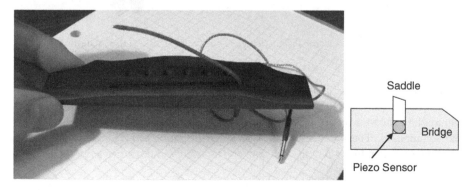

Fig. 4.52 Bridge with Under Saddle Pickup

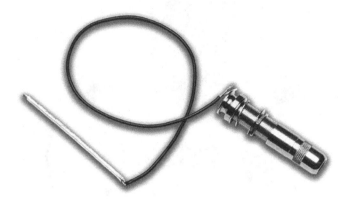

Fig. 4.53 An Under Saddle Piezo Pickup That Doesn't Require a Pre-Amp (Image courtesy of Duncan-Turner Acoustic Research, http://www.d-tar.com)

Fig. 4.54 Soundspot Piezoelectric Pickup (Image courtesy of Duncan-Turner Acoustic Research, http://www.d-tar.com)

by the velocity of the moving surface. Thus, it is often preferable to amplify an acoustic guitar using a sensor that directly senses the motion of the soundboard. Figure 4.54 shows a pair of piezoelectric sensors designed to be mounted directly to the soundboard. The two sensors are connected in parallel to a single output jack. Like the sensor in Fig. 4.53, this unit can also be used without a pre-amp.

4.9 Amplifiers

There certainly would be no electric guitars without amplifiers. However, it is sometimes difficult to understand what exactly they do. There is also a rich cultural tradition among guitarists about different amps that is often motivated by quite subjective evaluations (musicians sometimes jokingly refer to mojo). For example, there is a lively and ongoing debate about using vacuum tube amplifiers or solid state amplifiers (those that use transistors instead of vacuum tubes). This debate also touches on the benefits of different speakers, the ways different amplifiers tailor or distort sound and many other subtle things. The fundamental principles of amplifiers sometimes get lost in the background noise.

The most basic function of an amplifier is increase the amplitude (voltage) and power of an input signal. It is perhaps easier to understand this function in terms of familiar hardware. Imagine trying to drive a speaker directly with the output from a pickup and without an amplifier. The resulting sound level would be too low to hear.

The problem is clearly that the pickup alone doesn't make enough power to move the speaker. In electrical terms, power is simply the product of voltage and current, $P = i \times V$, and the unit of power is Watts. A small practice amp may be rated at 10 W and a large amp designed for stage performance might be rated at 250 W. Conversely, the pickup alone might produce milli-Watts, thousandths of a Watt, not enough to drive a speaker.

It is important to note that the power rating of an amplifier refers to electrical power, not acoustic power. For example, a 50 W amplifier doesn't produce anything like 50 acoustic Watts. Rather, this power rating refers to the electrical power that is dissipated by the amplifier.

In order to organize the discussion, we need to have some way of organizing the types of amplifiers. One way is to differentiate by the design characteristics and the other is to differentiate by the type of hardware used. Amplifier designs are designated with letters that describe what exactly they do. The most important classes of amplifiers for use with guitars are classes A, B and D [19]. The hardware distinction is simpler; all amplifiers are made using either transistors or vacuum tubes. Transistor-based amps are usually called solid state amps.

The letter designations for amplifiers are important to understand since they describe a fundamental operating principle. These classifications don't depend on whether the amplifier is made with solid state components or tubes. Any sound can be expressed as a collection of sine waves of different frequencies, so it makes sense to talk about how an amplifier handles a single sine wave. Class A amplifiers amplify an entire cycle of a sine wave as shown in Fig. 4.55. The transistor is used here in a common emitter configuration. Class A amplifiers generally have low noise and can be simple, but are inefficient. In fact, the theoretical efficiency limit for a Class A amp is 50%. Thus, it will dissipate at least one watt as heat for every watt delivered to the speaker.

Class B amplifiers amplify half the sine wave as shown in Fig. 4.56. This is an emitter follower configuration. Clearly, this amplifier significantly distorts the signal. The solution is to use two transistors (or tubes) so that each one can amplify

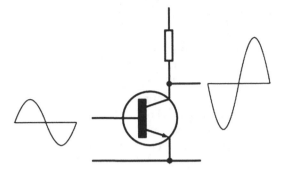

Fig. 4.55 Operation of a Class A Amplifier (Wikimedia Commons, image is in the public domain)

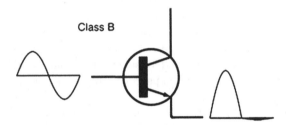

Fig. 4.56 Operation of a Class B Amplifier (Wikimedia commons, image is in the public domain)

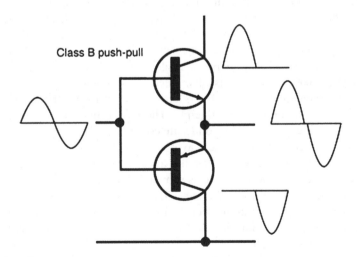

Fig. 4.57 Operation of a Class B Push-Pull Amplifier (Wikimedia commons, image is in the public domain)

half the input signal as shown in Fig. 4.57. This is generally called a push-pull amplifier. This approach makes the amplifier more efficient, with the theoretical limit being $\pi/4$ (78.54%). The complication, though, is that the two halves of the amplified signal may not join exactly back together and there may be distortion – called crossover distortion. This effect can be minimized with careful design, though sometimes at the cost of slightly decreased efficiency. Many guitar amplifiers are Class B push-pull designs.

A relatively recent development is digital amplifiers, called Class D. These are digital amplifiers in which the input signal is 'chopped' into segments that are then converted to a series of pulses.

A digital amplifier works by converting the analog signal from the guitar into a digital signal and then using that digital signal to drive a power supply that can switch on and off rapidly. This high power digital signal is sent to the speakers after the high frequency content has been removed by a low pass filter. The details are too complicated to be described completely here, but there are a few basic ideas worth illustrating.

One of the most important concepts is how to turn an analog signal into a digital one. There are several ways to encode information in a digital signal [20], but one often used for digital amplifiers is pulse width modulation (PWM). In this approach, the amplitude of the digital signal is always either 0 or V (where V is some positive voltage) and the information contained in the analog signal is encoded in the width of the pulses.

As you might expect, there is more than one way to create a pulse width modulated digital signal from an analog one. A simple one is to compare it with a sawtooth or triangular wave which serves as a reference signal. Whenever the analog input signal is larger than the reference signal, the resulting digital value is V– the high state. Otherwise, the digital value is 0 – the low state. This is called the intersective method.

Figure 4.58 shows two plots. The upper plot shows an analog signal and a sawtooth reference waveform. The lower plot shows the resulting PWM signal. The maximum pulse width is determined by the width of a single cycle in the sawtooth reference waveform. The width in this example was chosen to be 0.04 sec so that the reference wave would be easy to see. However, it is generally desirable for the period of the reference signal to be much shorter than the shortest period in the input signal. It is typical for the reference signal frequency (also called the switching frequency) in audio amplifiers to be hundreds of kHz. Note that the frequency is just the inverse of the period so that a 200 kHz frequency would correspond to a period of 5 μsec.

Figure 4.59 shows the basic layout of a class D amplifier. The PWM signal is used to drive a power supply capable of switching on and off rapidly. The result is a PWM signal similar to, but with much more power than the original one. Obviously, the last step is to convert the digital signal back to an analog signal that can drive the speaker. This is done using a lowpass filter.

Any signal can be decomposed into a collection of sine waves of differing frequencies and amplitudes - this is the basic idea behind the Fourier transform.

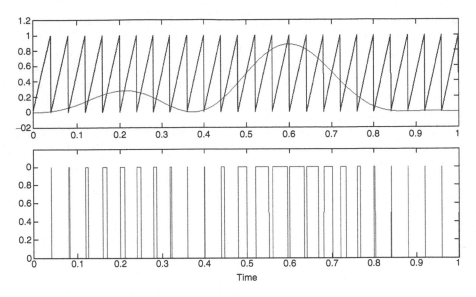

Fig. 4.58 Pulse Width Modulation of an Input Signal

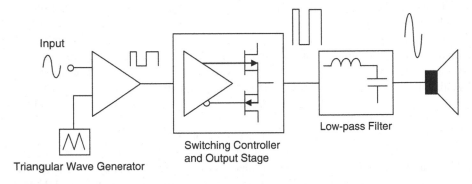

Fig. 4.59 Layout of a Class D amplifier (Redrawn from Wikimedia Image)

The very low frequency components of the PWM signal form the analog signal from which it was generated. The higher frequency components are what turn it into a pulsed signal. By filtering out the high frequency components of the high power PWM signal, one is left with a high power version of the analog input signal. Figure 4.60 shows the high frequency and low frequency components of the PWM signal from the bottom of Fig. 4.58.

The upper plot shows the low frequency components of the PWM signal superimposed over the original signal. It's barely possible to tell that there are two curves. The lower plot shows the high frequency components of the PWM signal and the original signal. In this case, the pulsed nature of the PWM signal is still very clear.

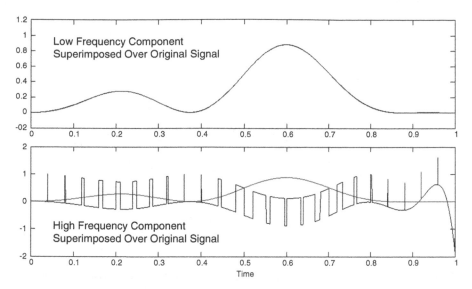

Fig. 4.60 Low and High Frequency Components of Sample PWM Signal

A key advantage of digital amplifiers is their efficiency. Since there is no need for large resistors, much less energy needs to be dissipated as heat. A Class D amplifier acts like a power supply that can be rapidly switched on and off. An ideal switch can conduct current with no voltage drop, so there is no power loss. The theoretical efficiency of a class D amplifier is 100% so that all the power supplied to it is delivered to the speaker. In practice, there are some losses, but measured efficiencies of 90% are common [21].

One last thing to note is that PWM is very different than another common scheme used in digital audio processing called Pulse Code Modulation (PCM). PCM is used to represent sampled analog signals in digital form. A PCM signal is just a digital representation of the value of the analog signal at discrete time steps. Figure 4.61 shows the PCM representation of a sine wave.

4.9.1 Solid State Amplifiers

The solid state amplifier was made possible by the invention of the transistor in 1948 and their commercial availability in the 1950s. By the late 1960s, they were cheap enough to be used in a range of consumer electronics including guitar amplifiers [22].

Solid state amplifiers have traditionally been made using individual components (transistors, resistors and capacitors, mostly) soldered to a printed circuit board. This approach is often referred to as using discrete components.

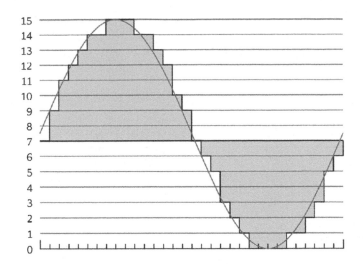

Fig. 4.61 PCM Representation of an Analog Signal (Wikimedia Commons, image is in the public domain)

Fig. 4.62 TL061 Op Amps Showing DIP-8 Packaging

However, the widespread availability of integrated circuits (ICs) has made designing with discrete components much less common. Integrated circuits have multiple components on a thin substrate of semiconductor material and are generally designed to be easy to use in practical circuits. ICs are in essentially all consumer electronics. The most familiar might be digital computer chips, but there are hundreds of analog IC chips useful for audio design. The most useful ICs for guitar amplifiers are operational amplifiers – usually just called op amps.

Fig. 4.63 A Tweezer Soldering Iron being Used on a Surface Mount Component (Wikimedia Commons, image is in the public domain)

An important practical consideration in choosing components is their packaging. Components can be mounted using through holes or can be surface mounted. Figure 4.62 shows two small ICs (TL061 op amps) and the eight pins that are inserted into matching holes in the circuit board. This particular packaging is called a DIP-8 (Dual Inline Package with eight pins). This type of mounting is convenient for prototyping using breadboards and for small production runs.

An alternative, now used in most production electronics, is surface mounting. Surface mounted devices (SMDs) are much smaller than equivalent through mounted components (components mounted in holes drilled through the circuit board), so the final product can be much smaller. Figure 4.63 shows a surface mounted component being placed using a tweezer soldering iron. While SMDs allow very compact circuit boards, they are difficult to place by hand and are generally installed using automated machinery.

Whether made using discrete components or ICs, solid state analog amplifiers are made using essentially four types of components: transistors, resistors, diodes and capacitors. However, transistors are at the core of any design.

The transistor is often described as a switch and that is correct, but it is not at all the whole story. It is roughly correct to think of a transistor as analogous to a valve that controls the flow of water in a pipe. It takes very little mechanical energy to open and close a valve, but that small amount of energy can control the flow of a large amount of water that, in turn, might have a large amount of kinetic or potential energy. In a similar way, a transistor uses a low power signal to control a larger one.

In order to understand the role of transistors, one must first sort out the terminology and a few basic concepts. There are essentially two types of transistors: bipolar junction transistors (BJT) and field effect transistors (FET). The physics of

Fig. 4.64 Two Types
of Bipolar Transistors

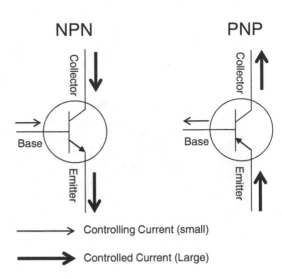

how they work are beyond this discussion; it is enough to know that they do basically the same thing, but have slightly different properties. FETs have superseded BJTs in digital ICs, but not in linear ICs. FETs are typically only used at the input stage of linear ICs. The remaining sections are mostly BJTs.

In the most basic terms, a BJT acts as a current-controlled current source while an FET acts as a voltage-controlled current source. That is, the BJT makes a current proportional to input current and the FET makes a current proportional to the input voltage. Also BJTs are normally off while FETs and depletion mode MOSFETs are normally on (MOS stands for Metal Oxide Semiconductor). If there is no signal input to the BJT, there is no amplified output. The MOSFETs used in digital ICs and power amplifiers are enhancement mode devices. These are normally off, like BJTs.

The BJT has three leads, called the collector, the emitter and the base. The signal to be amplified is connected to the base lead and the larger current – the one being regulated – moves between the collector and emitter.

There are two types of bipolar junction transistors, PNP and NPN. The names come from descriptions of the semiconductor materials inside the transistors. The practical difference between the two types is the way current flows through them. Figure 4.64 shows their basic operation.

Note that there are different types of FETs with the most common being the junction field effect transistor (JFET) and the metal oxide semiconductor field effect transistor (MOSFET). The terminals are called the gate, source and drain and are analogous to the base, collector and emitter on the BJT. The controlling voltage is applied between the gate and the source. Figure 4.65 shows circuit symbols for the JFET and one type of MOSFET (the depletion mode MOSFET).

Like most other discrete components, FETs are surface mounted in most commercial applications. Figure 4.66 shows a typical surface mount MOSFET.

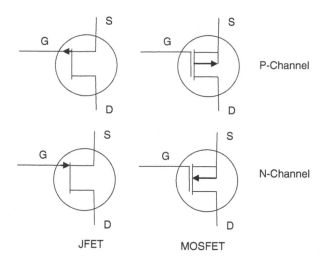

Fig. 4.65 Symbols for FET and MOSFET

Fig. 4.66 Surface Mount MOSFET (Wikimedia Commons, image is in the public domain)

In theory, a single transistor could be used as the basis for an amplifier. However, such a simple circuit is not usually practical. As a result, even the simplest practical transistor amplifiers generally have several transistors and supporting circuit runs. These additional components do things like set the gain value, ensure that input impedance is high (so the input signal is not affected) and output impedance is low (so that the output signal is not affected). They might also ensure the amplifier is stable under a range of different inputs. It is often simpler to use op amps in place of complicated arrangements of discrete components.

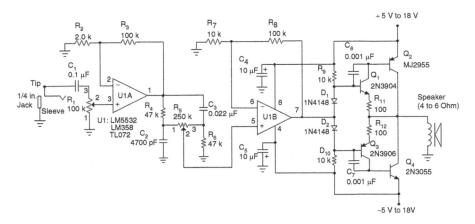

Fig. 4.67 Schematics for a Simple Solid State Guitar Amplifier (Image courtesy of Denny Dailey, Butler County Community College, http://www.BC3.edu)

Figure 4.67 shows a circuit diagram for a simple solid state amplifier designed by Denny Dailey, currently on the faculty of Butler County Community College. This circuit uses two op amps ahead of two power transistors which are paired to drive a speaker.

4.9.2   Tube Amplifiers

The precursor to the transistor was the vacuum tube, sometimes called a valve. Vacuum tubes were universally used in amplifiers before the transistor became available. Tube guitar amplifiers have a distinctive sound because of the way the signal distorts and this 'tube sound' is preferred by some guitarists over the sound produced by solid state amplifiers [23,24]. As a result, tube guitar amplifiers, along with a very limited selection of vacuum tubes, are still in production. Figure 4.68 shows a typical vacuum tube. This one is a 12AX7, a tube still in production and used for guitar amplifiers.

Audio amplifiers are the only consumer products that still use vacuum tubes and the reasons are easy to understand. Tubes are very large compared to discrete transistors and enormous compared to elements of an integrated circuit. Tubes also have very high power consumption because they need to heat an internal grid or plate. Tube amplifiers require large, heavy transformers and use internal voltages of hundreds of volts DC. In fact, people working on tube amplifiers should be very careful because of the danger of being shocked or even electrocuted.

It appears that tube amplifiers are an increasingly specialized product. While they are still being manufactured, the numbers are small and there are increasingly

Fig. 4.68 An Audio Vacuum Tube (Wikimedia Commons, image is in the public domain)

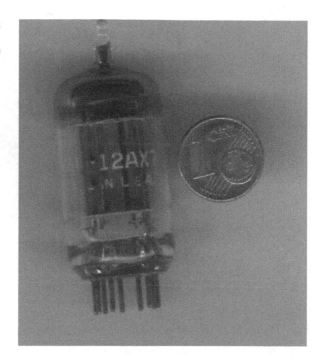

attractive alternatives that involve modeling the sound of a tube amp using either solid state hardware or software. For those who want a true tube circuit, it is possible to put a tube pre-amp in between the instrument and a solid state power amplifier. Figure 4.69 shows a tube pre-amp in a foot pedal package. This model uses two 6111 tubes – extremely small tubes that give the desired acoustic properties without requiring the large power supplies and heat dissipation associated with power amps.

Another approach to getting the sound of a tube amplifier without the power consumption of vacuum tubes, is to distort the signal using either a specially designed solid state analog circuit or software working with a digital circuit. For example, there are a number of effects pedals on the market that use a solid state circuit to simulate the effect of a tube amplifier on the signal coming from the instrument. There are also a number of amplifiers and effects pedals that use software to simulate tube amplifiers. These are generically called modeling amps or modeling boxes. They have a digital processor that modifies the input signal using algorithms that have been programmed in. Sometimes, these algorithms can be updated by the user. Figure 4.70 shows a small effects processor that includes software models of dozens of amplifiers. It includes a computer interface so that the user can update the software to add new features and also use it as a hardware interface for digital recording.

Fig. 4.69 A Foot Pedal Tube Preamp (Image courtesy of Seymour Duncan, seymourduncan.com)

Fig. 4.70 A Small Guitar Effects Processor That Includes Amplifier Models (Wikimedia Commons, image is in the public domain)

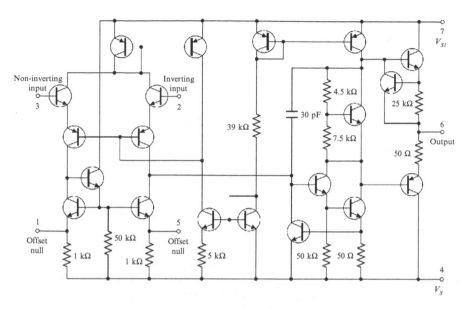

Fig. 4.71 The Circuit Diagram for a 741 Op Amp (Wikimedia Commons, image is in the public domain)

4.9.3 Preamps

A preamp is a low power amplifier that is generally placed between the guitar and either a power amplifier or headphones. It can have several different purposes, but it is generally used to increase the amplitude of the signal coming from the pickup and to match impedances between the pickup and another component. Preamps are generally built around a solid state operational amplifier, an op amp. Op amps are composed of a collection of many circuit elements all packaged in a small unit with metal connectors (see Fig. 4.62). They are designed to be used with a small number of resistors and capacitors that set the gain as well as the input and output impedance. For example, it is possible to set the output impedance of the pre-amp to be very low so that more of the output voltage can be dropped across the amplifier.

Op amps have alpha-numeric designations. One of the first commercially available designs and one that is still widely used is the 741. It was introduced by Fairchild Semiconductors in 1968 as the μA741. Since then, versions of this design have been produced by other manufacturers. Figure 4.71 shows the circuit diagram for the 741. Though still widely used, the 741 is not generally a good choice for audio applications.

Op amps are usually inexpensive and there are models designed specifically for audio use. They have flat frequency response in the human hearing range and draw very little current for good battery life. While it is possible for an op amp to be part

Fig. 4.72 Circuit Diagram
Symbol for an Op Amp

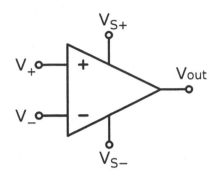

of a very simple circuit, practical considerations require some additional elements. The circuit diagram symbol for an op amp is shown in Fig. 4.72

Where V_+ is the non-inverting input, V_- is the inverting input, V_{S+} is the positive power supply, V_{S-} is the negative power supply and V_{out} is the output signal. Op amps are very versatile and extended works have been written about them. This discussion will be limited to three specific circuits. Note that the positive and negative power supply inputs are often not shown on circuit diagrams.

Pre-amps used for magnetic pickups basically do two things: increase the amplitude of the signal from the pickup (gain) and change the output impedance from the guitar. If a pre-amp is used to increase the amplitude of the signal from the pickup, it is possible to use fewer turns of wire in the coil(s). If there are fewer turns of wire, the pickup is necessarily less sensitive to electromagnetic noise and the result can be a 'cleaner' signal – one with a higher signal to noise ratio.

Three requirements to keep in mind when designing a simple pre-amp are: whether the amplifier should invert the signal or not; whether it will be powered with a battery or some other kind of power supply; and whether it needs to amplify a steady or a varying signal. For a guitar, the pre-amp will likely be powered by an on-board battery and will certainly need to amplify a signal that varies with time and that can change signs as it varies. This an alternating current signal or, more usually, an AC signal. The average value of AC signals is often zero and this is almost always the case for passive guitar pickups.

Since the pre-amp will be battery powered, $V_{S-} = 0$ and the result is called a single supply amplifier. Op-amps can only make an output signal such that $V_{S-} < V_{out} < V_{S+}$. For a single supply pre-amp, this would mean that any negative portions of the signal would be clipped at 0 Volts and only the positive portion of the signal would be output. The solution is to bias the input so that its average is something other than zero. If the pre-amp is to be powered with a 9 V battery, it makes sense to bias the input to 4.5 V.

Figure 4.73 shows a sine wave with a half amplitude of 1, that is the wave oscillates from -1 to +1. The center plot shows the same sine wave with the negative portion clipped. This is what would happen if this signal was input directly to an op amp with a single supply. The lower plot shows the original sine wave biased by 2 V, so the wave oscillates in the range from 1 – 3.

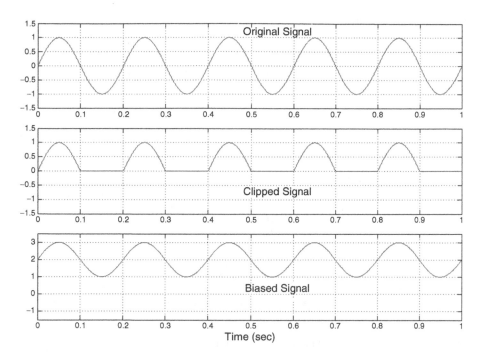

Fig. 4.73 A Sine Wave with Clipping and Bias

   Whether the op-amp should invert the signal or not often depends on whether the
signal from the pickup is going to be combined with signals from other sources.
If the output from the op amp isn't going to be combined with another signal (say,
from another pickup), it may not matter whether the op amp inverts the signal or
not; the change in the sound caused by inverting the signal is so minor that it
wouldn't be audible. Some acoustic guitar preamps have a phase switch that allows
the user to select inverting or non-inverting operation to help reduce the effects of
audio feedback. If the output of an inverting op amp is to be combined with another
signal that hasn't been inverted, the change may cause undesirable interactions
between the two.
   Figure 4.74 shows a signal that has been amplified by a factor of two. The top
plot shows the inverted output and the bottom plot shows the non-inverted output.
   In order to lay out the circuit for a simple pre-amp, we need to define
requirements. Clearly, it has to amplify an oscillating voltage – an AC signal.
Further, let's assume we don't want to invert the signal and that the pre-amp will
be powered with a battery. Figure 4.75 shows the schematic for a non-inverting,
single supply pre-amp [25].
   This circuit diagram is a fairly standard one with a few modifications. Part of the
utility of op amps is that gain and impedance are easy to adjust simply by changing

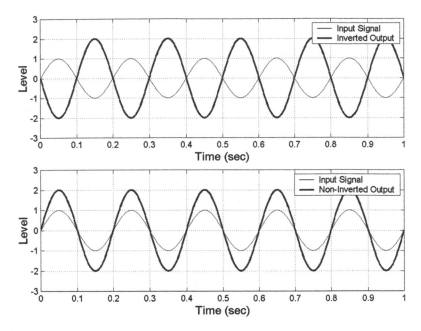

Fig. 4.74 Output from an Inverting and a Non-inverting Op Amp

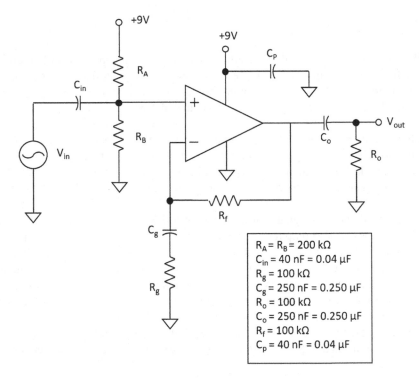

Fig. 4.75 Circuit Diagram for Non-Inverting Pre-Amp

the values of a few external capacitors and resistors. Gain is defined as the ratio of output to input, $A = V_{out}/V_{in}$. For this pre-amp,

$$A = \frac{V_{out}}{V_{in}} = 1 + \frac{R_f}{R_g} \tag{4.18}$$

The input impedance is

$$Z_{in} = \frac{R_A R_B}{R_A + R_B} \tag{4.19}$$

The input signal feeds into the non-inverting (+) input of the op amp. However, there is a capacitor and two resistors on this portion of the circuit. The capacitor is there to remove low frequency components of the input signal, including the DC component. This ensures that the signal doesn't have any bias (it has a mean value of 0 V). Remember the expression for the impedance of a capacitor, $Z_C = 1/j\omega C$. This expression says that Z_C becomes large as the frequency decreases. Thus, a capacitor acts like an open circuit when $\omega = 0$ so the DC component of the input signal cannot pass.

Note also that the resistors R_A and R_B form a voltage divider. The bias voltage, V_B, that will be added to the input voltage is

$$V_B = \frac{R_B}{R_A + R_B} 9V \tag{4.20}$$

If $R_A = R_B$, then bias voltage will be half the voltage from the battery. We want the bias voltage to be 4.5 V and the input impedance to be at least 10 times higher than the output impedance of the pickup. For a typical magnetic pickup, it is desirable that $Z_{in} \approx 100\,k\Omega$. From Equation 4.19, if $R = R_A = R_B$, then $Z_{in} = R/2$. Thus, we can get the desired bias voltage and the desired input impedance if $R_A = R_B = 200\,k\Omega$.

The next parameter that needs to be set is the value of the input capacitor, C_{in}. The combination of C_{in}, R_A and R_B forms a highpass filter. A high pass filter passes only signals with frequencies higher than f_c, the cutoff frequency, where

$$f_c = \frac{1}{2\pi RC} \tag{4.21}$$

At first glance, there would seem to be no need for a guitar amplifier to amplify frequencies less than 82.4 Hz since that is the fundamental frequency of the low E string. However, there are situations in which a player might want to tune the instrument lower than normal. If the cutoff frequency is to be 40 Hz (251.3 rad/sec), then $RC \approx 0.004$. Figure 4.76 shows the gain (often called a transfer function) for the resulting high pass filter. Note that the gain is not exactly 1 immediately above the cutoff frequency. Rather, the gain curve blends smoothly from the straight section below the cutoff frequency to the horizontal section above the cutoff frequency.

A battery acts as ground for an AC signal, so R_A and R_B can be considered to be in parallel. If $R_A = R_B = 200\,k\Omega$, then the equivalent resistance for the two of them in parallel is $100\,k\Omega$. Thus $C_{in} = 40$ nF. Note that C_g and R_g form another

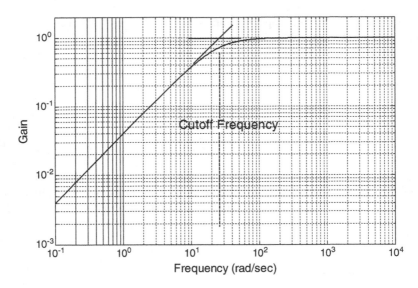

Fig. 4.76 Transfer Function for High Pass Filter

highpass filter as do C_o and R_o. In order to reduce the chance of coupling between these three highpass filters, the corner frequencies of the one going to ground and the one on output should be set lower. A common guideline is that the corner frequencies should be around 1/7 of the value of the input filter. If $R_g = 100\,k\Omega$ and $C_g = 250\,nF$, then the corner frequency is about 6.4 Hz. These values are also used for C_o and R_o.

The last thing to specify is the type of op amp. A quick survey of catalogs from electronics suppliers shows that there are many different op amps readily available and it's easy to get overwhelmed by the number of choices. As a practical matter, many designers select a small number of designs that meet their most common needs, using other designs only when necessary.

The TL061/062 series of op amps is a good choice for pre-amps. They have very low noise, draw little power and allow the designer to select whatever gain is most suitable for the application. Some features can be added to the basic preamp design to tailor it for use in a guitar. Figure 4.77 shows a nice, yet simple pre-amp circuit provided by Kevin Beller of Seymour Duncan. It includes a few more components than the circuit in Fig. 4.75.

Components R1 and C1 form a low pass filter with a corner frequency of about 72,300 Hz. This is far above the human hearing range and prevents high frequency electromagnetic noise from reaching the op amp. R2 drains the charge from C2 so there is no click when the jack is plugged in and C3 blocks DC voltage. R3 and R4 form a voltage divider to bias the input voltage. This allows the op amp to use a single supply (like a 9 V battery). R5 is a bias resistor; R5 in parallel with R2 sets the input impedance. R6 is a feedback resistor that, along with R7 sets the gain. With R6 = R7, the pre-amp gain is 2. C4 prevents the circuit from having gain at very low frequencies and limits DC offset in the output signal. C5 and R9 form a

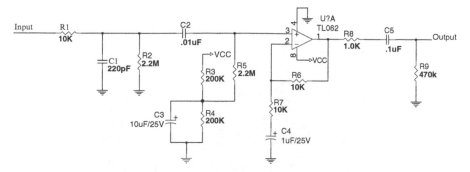

Fig. 4.77 Practical Op-amp Circuit Using TL061/062 (Image courtesy of Kevin Beller, Seymour Duncan, http://www.seymourduncan.com)

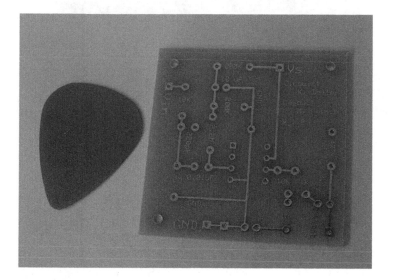

Fig. 4.78 Preamp Circuit Board without Components in Place

high pass filter with a corner frequency of about 3.4 Hz, eliminating any DC offset in the output signal.

Such a simple circuit is easy to implement using a printed circuit board. Figure 4.78 shows a board without the components in place. It's a little larger than a guitar pick and could probably be made smaller. Of course, using surface mount components would make it much smaller.

Figure 4.79 shows the same board with the components in place. They are well-spaced, so assembly is easy.

An alternative is to select an op amp designed specifically for minimum external parts count. The LM386 by National Instruments is just such a chip. It is designed for a single supply so it is easy to power with a battery. It also has a default internal gain of 20 that requires very few external components. The minimum gain is 20 and it can

Fig. 4.79 TL061 Preamp with Components in Place

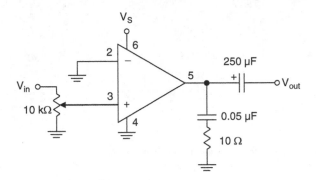

Fig. 4.80 Schematic for a Simple Pre-Amp Using an LM386 Op Amp

develop a maximum gain of 200 with the addition of a few additional external components. Power requirements are low and it has relatively little distortion – 0.2% with a 6 V supply and an 8 Ω load. Not surprisingly, it is a popular op amp for simple audio pre-amps. Figure 4.80 shows the schematic for a pre-amp with the default gain of 20 [26]. This pre-amp is a good choice for driving a small speaker or even a larger one if it is efficient. The low power – about ½ watt depending the version of the chip used – is enough for practicing. It can also drive headphones or ear buds, though a resistor might be needed to lower the volume. Note that power consumption is higher than for the TL061 circuit above; if powered by a 9 V battery, it will need to be changed more often.

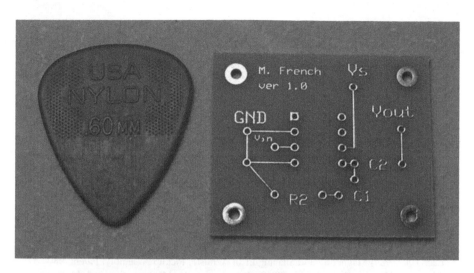

Fig. 4.81 A Printed Circuit Board for a Pre-Amp Using an LM386 Op Amp

Fig. 4.82 A Simple Pre-Amp Made Using a Custom Printed Circuit Board and an LM386 Op Amp

Figure 4.81 shows the first version of a printed circuit board laid out by the author to use the LM386. The circuit board is hardly bigger than a guitar pick, so finding room for it in the instrument is easy.

Figure 4.82 shows the same board populated. The largest component is the capacitor that filters the DC components from the output signal.

References

1. Hunter D (2008) The Guitar Pickup Handbook, Backbeat Books, New York.
2. Hembree G (2007) Gibson Guitars: Ted McCarty's Golden Era: 1948-1966, GH Books, Austin TX.
3. Reese RL (2000) University Physics, Brooks/Cole Publishing, Pacific Grove CA.
4. Lover SE (1959) Magnetic Pickup for Stringed Musical Instrument, Patent 2,896,491.
5. Self D (2010) Small Signal Audio Design, Focal Press.
6. Schultz ME (2007) Grob's Basic Electronics, McGraw Hill.
7. Nahin PJ (2010) An Imaginary Tale: The Story of I, Princeton University Press.
8. Beller K (2010) Private Correspondence.
9. Hartman WM (1997) Sound, Signals and Sensation, American Institute of Physics
10. Livingston JD (1997) Driving Force: The Natural Magic of Magnets, Harvard University Press
11. Coey JMD (2010) Magnetism and Magnetic Materials, Cambridge University Press
12. Beller K (2010) Purdue Guitar Workshop Presentation: Basics of Pickup Design
13. Fleisch D (2008) A Student's Guide to Maxwell's Equations, Cambridge University Press.
14. Hunter D (2008) The Guitar Pickup Handbook, Backbeat books
15. Lace DA (1989) Magnetic Field Shaping in an Acoustic Pickup Assembly, Patent 4,809,578
16. Alciatore DG and Histand MB (2012) Introduction to Mechatronics and Measurement Systems, 4th ed., McGraw Hill.
17. Katzir S (2006) The Beginnings of Piezoelectricity: A Study in Mundane Physics, Springer.
18. Jones DJ, Prasad SE and Wallace JB (1996) Piezoelectric Materials and Their Applications, Key Engineering Materials, Vols. 122–124.
19. Jacob M (2002) Power Electronics: Principles and Applications, Delmar
20. Holmes DG and Lipo TA (2003) Pulse Width Modulation for Power Converters: Principles and Practice, Wiley-IEEE Press
21. Self D (2009) Audio Power Amplifier Design Handbook, 5th ed, Focal Press.
22. Riordan M and Hoddeson L (1998) Crystal Fire: The Invention of the Transistor and the Birth of the Information Age, W.W. Norton & Co.
23. Barbour E (1998) The Cool Sound of Tubes, IEEE Spectrum, August, pp24–35.
24. Weber G (2004) Tube Guitar Amplifier Essentials, Kendrick Books
25. Kitchin C (2002) Demystifying Single Supply Op Amp Design, EDN Magazine, March 21, pp 83-90.
26. LM386 Low Voltage Audio Power Amplifier Data Sheet (2000) National Semiconductor

Chapter 5
Sound Quality

*French's Law: For every expert, there is an equal
and opposite expert*

There is no topic in guitar design more contentious or supported by more ambiguous data than sound quality. The quality of the sound produced by a guitar, often called tone or tonal quality, is obviously one of its most important qualities, but it may be the thing we know least about.

Luthiers often appear to have an instinctive or idiomatic understanding of what they can do to their own instruments in order to improve tone. However, they often have a difficult time explaining it in a precise way and, more particularly, in a way that accurately describes the underlying physics. For example, a builder might say that decreasing stiffness of a brace causes the sound to 'open up'. The effect may be real, but the description is not very helpful in understanding it.

In order to make sense of the problem of sound quality, it is greatly helpful if there is some way to organize the conversation. There are many possible organizational structures for a description of sound quality. However, the one here will start with pitch correction, a topic applicable to all guitars, then progress to topics specifically for electric guitars and, finally, to topics specifically for acoustic guitars.

5.1 Achieving Correct Pitch

The discussion in Chapter 2 outlined the way frets are spaced to give the correct pitch. Stretching and inharmonicity combine to slightly change the actual pitch so that the note played is almost never exactly correct. If bridge intonation is set using notes at the 12th fret, the notes sounded at each position will only be correct for the open strings and the strings being played at the 12th fret. The net effect is that the guitar is rarely really in tune.

R.M. French, *Technology of the Guitar*, DOI 10.1007/978-1-4614-1921-1_5,
© Springer Science+Business Media New York 2012

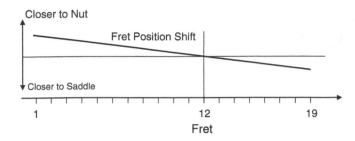

Fig. 5.1 Qualitative Representation of Fret Position Shift Calculated by Gilbert (1984)

One assumption in bridge compensation is that the resulting errors will be small enough to be negligible. For many players, this is indeed correct. However, some players are sensitive enough to notice even the small remaining pitch errors and there have been some attempts to more precisely correct the pitch of guitar strings by changing fret locations, modifying the nut, modifying the saddle or combinations of these. These methods should work for any fretted instrument, including basses and mandolins.

Before examining the basics of pitch correction, it is appropriate to survey some of the notable work in this area. One of the earliest articles on the subject of pitch correction in guitars is by W. Bartolini and P.A. Bartolini [1]. They focused on nylon string guitars and most of the article covers a detailed study of the dynamics of the bodies and their response to changes in mass. At the end of the article, however, they calculate the effect on pitch of strings stretching as they are fretted.

They make some reasonable assumptions combined with measurements of axial stiffness of several strings and calculate fret locations that vary slightly from the ones calculated using the nominal relationship based on the factor $\sqrt[12]{2}$. Importantly, they also describe a nut with slight end corrections for each string similar to those at the saddle.

Another article appearing only slightly later is by John Gilbert [2], who also addresses nylon stringed guitars. He makes assumptions similar to those by the Bartolinis, though slightly simplified, and arrives at similar conclusions. Rather than specifying individual corrections at the nut, he suggests a simpler approach of trimming the nut end of the fretboard. He considers an instrument with a nominal scale length of 25.70 in (653 mm) and arrives at a fret spacing scheme that shifts spacing so that the correction varies roughly linearly down the fretboard. Qualitatively, the correction follows the pattern shown in Fig. 5.1

Greg Byers did a more sophisticated analysis that accounts for both stretching and inharmonicity due to bending stiffness [3], though still limiting his presentation to nylon strings. His analysis of stretching is similar to that used by the Bartolinis and adds an additional correction for inharmonicity using equation 2.10. He calculates both nut and saddle corrections that vary from string to string and roughly agree with the results of others.

More recently several people have extended previous work to instruments using steel strings. Varieschi and Gower [4] did an analytical and experimental study that includes an analytical compensation model. Also, Elmendorp [5] did an analysis using simplified geometry and correlated the calculated results with measured ones.

Fig. 5.2 Headstock of an Acoustic Guitar Showing Compensated Nut (Image courtesy of Trevor Gore, http://www.goreguitars.com.au)

Trevor Gore, an Australian luthier with an extensive technical education, has developed a nut and saddle compensation method that is qualitatively similar to some of the others in the literature. He produces guitars with compensated nuts and saddles along with a number of other design features that are not necessarily visible [6] from the outside. Figure 5.2 shows the headstock of a Gore guitar with the compensated nut and Fig. 5.3 shows the accompanying compensated saddle. Note also the dark line showing in the bridge. This is a graphite lamination that increases the stiffness of the bridge and allows reduced bridge mass.

Another recent effort to calculate corrected fret positions along with nut and saddle compensation is by Gary Magliari and implemented on mandolins by Don MacRostie [7]. This approach calculates how much the strings stretch when fretted and uses three different means of adjusting the resulting frequencies: nut compensation, saddle compensation and fret spacing.

What follows is a summary of their results for a 25.5 in (648 mm) scale length using a popular set of steel guitar strings (D'Addario EJ17 medium acoustic guitar strings). String dimensions and published tension values [8] are shown in Table 5.1. Tension in Newtons was added by the author.

Axial stiffness (resistance to stretching) was measured using a simple test fixture and the six experimental stiffness numbers are used to calculate string frequencies. The ideal string equation (Equation 2.3) also requires mass per unit length. Magliari used a test fixture to measure tension required to bring the strings to the required pitch. This information was then used to calculate mass per unit length.

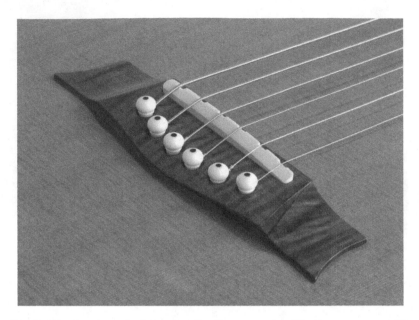

Fig. 5.3 Saddle of an Acoustic Guitar Showing Compensated Saddle (Image courtesy of Trevor Gore, http://www.goreguitars.com.au)

Table 5.1 Properties of EJ17 String Set

Note	Frequency Hz	Type	Diameter inches	mm	Tension lb	kg	N
E4	329.6	Plain	0.013	0.33	27.4	12.43	121.9
B3	246.7	Plain	0.017	0.43	26.3	11.93	117.0
G3	196.0	Wound	0.026	0.66	35.3	16.01	157.0
D3	146.8	Wound	0.035	0.89	36.8	16.69	163.7
A2	110.0	Wound	0.045	1.14	34.0	15.42	151.2
E2	82.41	Wound	0.056	1.42	29.0	13.15	129.0

Figure 5.4 shows the calculated frequency error with no compensation at all. Frequency error is expressed in cents, a common unit for expressing frequency changes or musical intervals much smaller than a half step. There are 100 cents in a half step and electronic tuners sometimes display frequency errors in cents. Where the ratio of two frequencies separated by a half step is 1.059463, the ratio of two frequencies separated by a single cent is 1.00057779. Thus, $1.00057779^{100} = 1.059463$.

Note that the largest frequency error occurs in the strings with the largest diameter of the load-carrying wire, B and low E. The B string is not wound, so the load-carrying wire is 0.017 in (0.43 mm). The low E string is wound, but is the largest string and has a largest core wire among the wound strings.

Conversely, the lowest frequency error occurs in strings with the smallest load-carrying wires, G and high E. The high E string is not wound, so the load-carrying

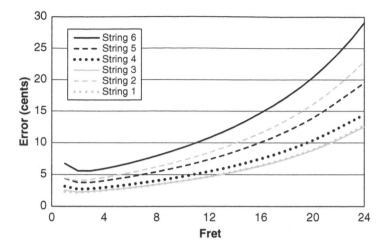

Fig. 5.4 Calculated Intonation Profile without Compensation (Data courtesy of Gary Magliari)

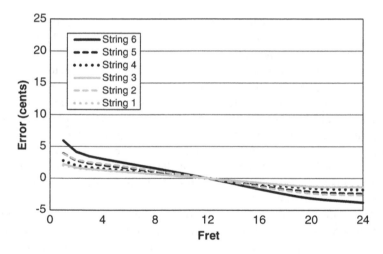

Fig. 5.5 Calculated Intonation Profile with Saddle Compensation (Data courtesy of Gary Magliari)

wire is 0.013 in (0.33 mm). The G string is the smallest wound string and has the smallest core wire of the wound strings.

The standard approach for pitch correction is to move the saddle slightly farther from the nut, a process usually called intonation or compensation. Figure 5.5 shows the calculated frequency errors for the six strings with saddle compensation. Typically, the saddle positions are set using the pitch at the 12th fret and this is reflected here. The result is that open string pitch is correct as is the pitch at the 12th fret. However, there is a slight residual error at every other fret.

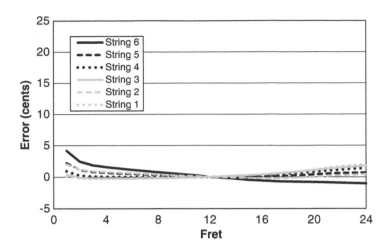

Fig. 5.6 Calculated Intonation Profile with Saddle Compensation and Shifted Nut Position (Data courtesy of Gary Magliari)

In practice, this error is small; not all players can even hear a frequency error of 5 cents. However, some can and there is no compelling reason to tolerate a known error if the means are at hand to correct it.

The next level of sophistication in correcting frequency errors is to allow the nut to shift one way or the other. Some builders routinely trim a small amount from the nut end of the fretboard, placing the nut slightly closer to the first fret than the ideal string equation specifies. Nut offsets on the order $0.010" - 0.030"$ (.254 mm $- 0.762$ mm) appear to be common among builders who use this technique.

Figure 5.6 shows the calculated frequency error resulting from shifting the nut slightly towards the first fret. Note that the predicted error is now less than less than 5 cents all the way down the fretboard and the largest error is only on the low E string at the first fret.

A further refinement is to individually compensate the strings at both the nut and saddle. Figure 5.7 shows the predicted frequency errors. With this additional level of correction, the frequency errors are confined almost exclusively to the higher frets. Error at the highest frets is actually higher than the previous case in which the nut was moved equally for all strings. The largest errors are above the 16th fret, hardly the busiest part of the fretboard for most players. Indeed, many acoustic guitars don't have more than 20 frets. This seems a good trade in exchange for nearly zero error at frets 1-16.

If the player spends most of the time between the nut and the 16th fret then compensating the nut individually for each string is probably the better choice. Since many players make relatively little use of the highest frets, there is a potential market for aftermarket compensated nuts and there are several companies offering them [9,10]. Figure 5.8 shows a commercially available compensated nut fitted to a Stratocaster.

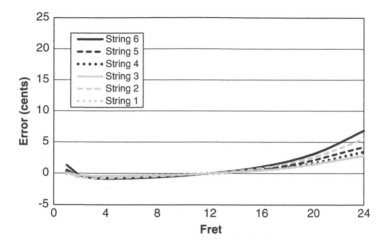

Fig. 5.7 Calculated Intonation Profile with Nut and Saddle Compensation (Data courtesy of Gary Magliari)

Fig. 5.8 A Commercially Available Compensated Nut on a Stratocaster (Image courtesy of Earvana, http://www.earvana.com)

Finally, with the mathematical tools in hand to calculate the effect of moving the locations of individual frets, it certainly makes sense to look for a way to slightly shift the fret positions to further reduce frequency errors.

One problem with this approach is that each string can require a slightly different fret spacing in order to eliminate predicted frequency errors. Magliari addressed the

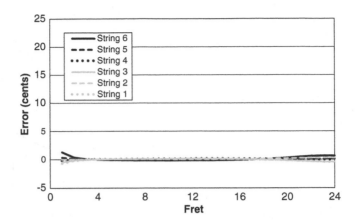

Fig. 5.9 Calculated Intonation Profile with Nut, Saddle and Fret Compensation (Data courtesy of Gary Magliari)

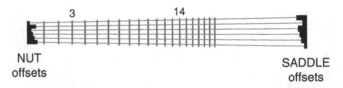

Fig. 5.10 Tension Compensated Fret Placement with Nut and Saddle Offsets (Image courtesy of Gary Magliari)

problem by slightly shifting the positions of each string and slightly changing the scale length for each string.

He calculated the adjusted fret spacing for each string independently, and shifted the patterns so that they lined up at the 3rd fret. He then adjusted scale lengths slightly so that also lined up at the 14th fret. Remember that the adjustments are on the order of 1 mm (0.039 in), so they are still about the magnitude of the scale length changes required by traditional saddle intonation.

While the 3rd and 14th frets could be straight, the remaining fret locations slightly disagreed. He then averaged the remaining fret locations so that all frets were straight. The resulting predicted frequency errors are very small, as shown in Fig. 5.9. At no point on the fretboard is the calculated error more than a few cents and it is essentially zero from frets 2 – 20.

Figure 5.10 shows an expanded view of the resulting compensation at the nut and saddle.

Rather than try to duplicate the calculations using the different methods presented here, it is enough to show a simplified, representative example with a plain string. Readers wanting to see more detailed calculations should go back to the original sources.

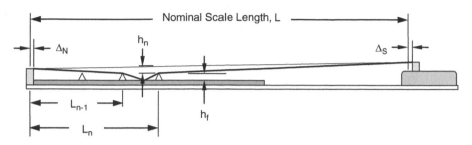

Fig. 5.11 Dimensions of a String Being Fretted

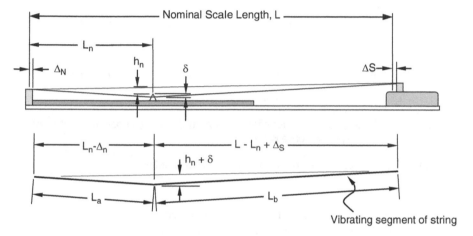

Vibrating segment of string

Fig. 5.12 Dimensions of Simplified String Geometry

To account for frequency changes due to stretching of the string, there needs to be a clear way of identifying the various dimensions for a string being fretted. Figure 5.11 shows the dimensions of a stretched string and fretboard with few simplifications. Note that there are offsets defined at both the nut and saddle. Also, the string is assumed to be pushed to the fretboard so there is additional stretching between the fret being played and the one above it.

Let's make a few simplifications to keep the analysis from becoming too cumbersome. This will slightly change the results, but the basic concept will remain. Figure 5.12 shows the simplified geometry, including the string in isolation. Rather than accounting directly for the additional tension due to the additional displacement between fret n and fret n + 1, let's just add a bit of additional displacement, δ. Of course, this reduces the accuracy, but the result is still qualitatively correct.

The analysis proceeds like this:

1. Find the tension required to bring the open string to the correct pitch
2. Find the length of stretched string and how much longer it is than the un-stretched string

Scale Length: $L := 25.5 \cdot in$ $L = 647.7\,mm$

Frequency: $f := 196 \cdot Hz$

Diameter: $d := 0.017 \cdot in$ $d = 0.432\,mm$

Density of Steel: $\rho_{steel} := 7850 \dfrac{kg}{m^3}$ $\rho_{steel} = 0.284 \dfrac{lb}{in^3}$

Find Area of String: $a := \dfrac{\pi}{4} \cdot d^2 = 1.464 \times 10^{-7}\,m^2$

Find Running Mass: $\rho := \rho_{steel} \cdot a = 1.15 \times 10^{-3}\dfrac{kg}{m}$ $\rho = 6.437 \times 10^{-5}\dfrac{lb}{in}$

Calculate Tension: $T := 4 \cdot f^2 L^2 \cdot \rho = 74.105\,N$ $T = 16.659\,lbf$

Note: Mathcad identifies lb as pound mass (a specious unit at best) pound force is lbf

Fig. 5.13 Mathcad Calculation of Tension for an Unwound G String

3. Find the change in frequency due to being stretched
4. Find the frequency change for each fret position
5. Add bridge compensation and find new frequency changes
6. Add nut compensation and re-calculate frequencies
7. Add fret compensation and make final frequency calculation

For this example, let's consider a plain steel G string for an electric guitar. The diameter of the string is 0.017 in (0.43 mm), the scale length, L, is 25.5 in (648 mm) and the published tension [8] is 16.6 lb (7.53 kg or 73.9 N). Figure 5.13 shows the calculation for the running mass using Mathcad, what is, at this writing anyway, a popular and powerful technical calculation program. Tension is calculated as well, just as a check to make sure the rest of the calculation makes sense. Note that Mathcad automatically converts units as necessary, so quantities can be initially defined in whatever units are convenient.

Following the outline above, let's start by finding the length of the stretched string

$$L_s = L_a + L_b = \sqrt{(h_n + \delta)^2 + (L_n - \Delta_N)^2} + \sqrt{(h_n + \delta)^2 + (L - L_n + \Delta_s)^2}$$

$$(5.1)$$

The change in length of the string is simply, $\Delta L = L_s - (L - \Delta_N + \Delta_s)$, so

$$\Delta L = \sqrt{(h_n + \delta)^2 + (L_n - \Delta_N)^2} + \sqrt{(h_n + \delta)^2 + (L - L_n + \Delta_s)^2} \tag{5.2}$$
$$- (L - \Delta_N + \Delta_s)$$

In order to calculate actual numbers, we need to have a definition for h_n. The distance from the string to the top of the frets varies along the neck. The height varies linearly along the neck, so it's easy to write a relationship between height and distance along the neck. If string height at the first fret is 0.020 in (0.508 mm) and 0.080 in (2.03 mm) at the 12th fret, then

$$h(x) = h_1 + \frac{h_{12} - h_1}{L_{12} - L_1}(x - L_1) \tag{5.3}$$

Where h_1 and h_{12} are the string heights at the 1st and 12th frets respectively. Let's also assume that $\delta = 0.020$ in (0.508 mm).

Next, there needs to be an expression relating a change in length to a change in tension.

$$\Delta T_n = \frac{AE}{L - \Delta_N + \Delta_s} \Delta L_n \tag{5.4}$$

The string is plucked after being fretted, so we can use the ideal string equation to calculate the new frequency. We're only looking at the change in frequency due to the string being stretched; we aren't worrying about frequency changes due to bending stiffness (inharmonicity).

$$r^n f + \Delta f_n = \frac{1}{2(L - L_n + \Delta_s)} \sqrt{\frac{T + \Delta T_n}{\rho}} \tag{5.5}$$

Where $r^n f$ is the ideal frequency at each fret. The frequency error is small compared to the frequency, so it's more informative to plot a ratio of the calculated frequencies to the ideal frequencies.

$$R = \frac{r^n f + \Delta f_n}{r^n f} \tag{5.6}$$

Let's start by looking at the frequency ratios for a string with no bridge or nut compensation. Figure 5.14 shows the calculated frequency ratios. The maximum

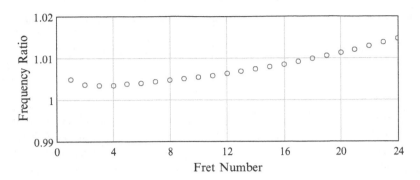

Fig. 5.14   Frequency Error with No Compensation

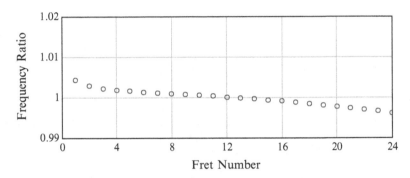

Fig. 5.15   Frequency Error with Saddle Compensation, $\Delta_S = 0.155$ in $= 3.94$ mm

error is about 1.5% at the 24th fret. Perhaps more significantly, the string is predicted to be more than 0.5% sharp over most of the fret board.

As the next step, Fig. 5.15 shows the effect of adding compensation at the saddle. The compensation is chosen so that the frequency ratio at the 12th fret is 1.00. Shifting the saddle back 0.155 in (3.94 mm) reduces the predicted frequency error at the 12th fret to essentially zero and to less than 0.5% over the entire fretboard. An interesting change is that the shape of the curve has changed; it had been increasing after the 4th fret and the saddle compensation has changed it so that it decreases after the 4th fret.

Finally, nut compensation is added to further reduce the predicted frequency errors. Roughly speaking, saddle compensation rotates the error curve and nut compensation shifts the curve vertically. Since both nut and saddle adjustments are now possible, Δ_S is reduced so that the majority of the error curve is approximately horizontal and Δ_N is set to shift the entire curve down so that there is no error at the 12th fret. The result, shown in Fig. 5.16, is that the largest calculated error is 0.2% at the first fret, and there is essentially no error above the second fret.

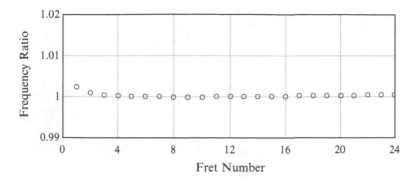

Fig. 5.16 Frequency Error with Nut and Saddle Compensation, $\Delta_N = 0.06$ in $= 1.52$ mm, $\Delta_S = 0.10$ in $= 2.54$ mm

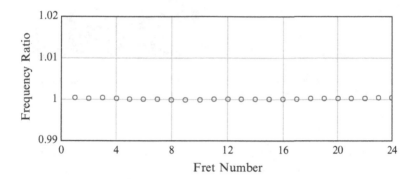

Fig. 5.17 Frequency Error with Nut and Saddle Compensation and First Two Fret Locations Shifted

An interesting and potentially very useful approximation for nut and saddle compensation has been suggested by Trevor Gore [11]. A builder not wanting to go through the analysis can get a noticeable improvement by taking the bridge compensation for each string, applying half that value to the nut and half to the saddle.

To this point, there has been no attempt to shift the fret positions. It is relatively easy to compensate the strings individual strings at both the nut and the saddle. Shifting the frets is more difficult since each string might require different fret positions. That said, shifting the positions of the first two frets slightly closer to the nut essentially eliminates predicted frequency error for this string. Figure 5.17 shows the predicted frequency errors when the first fret is shifted forward by 0.050 in (1.27 mm) and the second fret is shifted forward by 0.015 in (0.381 m).

Finally, there has been no attempt to account for string bending stiffness by using a real string equation like Equation 2.10. Like stretching, bending stiffness will also slightly increase the actual frequency of the string.

Fig. 5.18 A Mandolin Being Modified to Use a Compensated Fret Spacing (Image courtesy of Don MacRostie, http://www.reddiamondmandolins.com)

The system developed by Gary Magliari has been implemented by Don MacRostie of Red Diamond Mandolins. He has not only used it on his own instruments, but also retro-fitted it to existing instruments. Figure 5.18 shows a mandolin in a specially made fixture on a CNC router at Red Diamond. The existing frets are removed, then the fret slots are cleaned out with a very fine tapered mill and the filled with thin ebony strips. Then, a new set of fret slots is cut. For obvious reasons, this approach requires great confidence in the equipment and steady nerves.

There is at least one commercially available system of compensating frets along the neck of a guitar. True Temperament in Stockholm, Sweden [12] offers several necks based on four different temperament systems. One is the familiar equal tempered scale used on almost all guitars. Rather than making small approximations to allow the frets to be straight, this approach uses bent frets cast from silicon bronze. Figure 5.19 shows an electric guitar with this fretting system. The curved frets allow compensation to be tailored for each individual string. Of course, this comes at the price of having to cut curved fret slots and to cast matching frets.

The ability to manufacture and install curved frets offers the option of exploring other tempering systems. An interesting variation on the equal tempering system uses a modern interpretation of well tempering as used by J.S. Bach in The Well Tempered Clavier [13] as shown in Fig. 5.20.

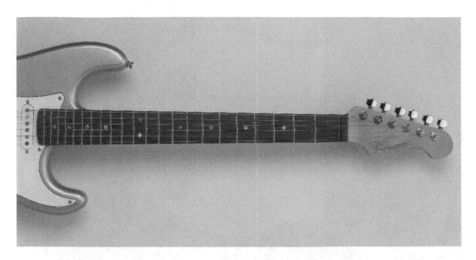

Fig. 5.19 An Electric Guitar Neck with Frets Individually Compensated for Each String (Image courtesy of True Temperament AB, http://www.truetemperament.com)

Fig. 5.20 An Acoustic Guitar Made to Use Well Tempering (Image courtesy of True Temperament AB, http://www.truetemperament.com, guitar by Sanden Guitars, http://www.sandenguitars. com)

5.2 Effect of Pickup Position on Tone

The tonal characteristics of an electric guitar are primarily determined by the electronics. Materials do play a role, but it is minor in comparison. The effect of pickups on electric guitar sound is essentially due to two things, the type of pickups and their locations. The effect on tone of pickup design is discussed in Chapter 4. Now, let's look at the effect of pickup position.

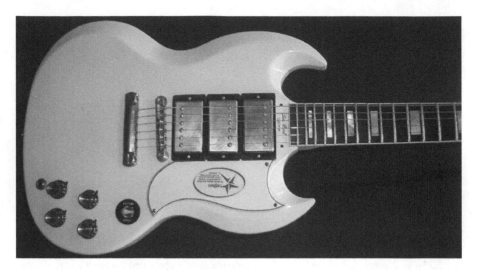

Fig. 5.21 A Gibson SG Custom Electric Guitar with Three Pickups (Wikimedia Commons, image is in the public domain)

It's typical for an electric guitar to have two or three pickups along with a switch to select between them. Figure 5.21 shows a Gibson SG Custom with three humbuckers.

Having three pickups adds complication, weight and expense, so there must be a good reason that so many electric guitars have multiple pickups. It can't be volume since that is easily taken care of by the amplifier. Rather, the different pickups offer different tonal choices for the player.

Put in the simplest terms, the closer to the bridge the pickup is placed, the more high frequency content will be in the resulting signal. The bridge pickup is sometimes called the lead pickup because lead players, particularly when playing solo passages, tend to prefer the bright, treble-laden tone from the bridge pickup. Conversely, the neck pickup is sometimes called the rhythm pickup and produces a much softer tone with less high frequency content.

The same type of effect can be produced by simply plucking the guitar at different positions. If the strings are played very near the bridge, the tone is bright, with lots of treble (high frequency content). If the strings are played farther from the bridge, the tone is smoother, with more low frequency content. The difference is explained by the basic physics of vibrating strings.

As explained in an earlier chapter, strings don't vibrate at random frequencies; if they did, they wouldn't be useful for making music. Rather, they vibrate at a fundamental frequency and (approximately) integer multiples of the fundamental. Each of these resonant frequencies is accompanied by a unique shape called a mode shape. If a string were to be excited at exactly one of its resonant frequencies, it would vibrate with only one mode shape. If the string was also illuminated by a strobe light to freeze it at upper limit of its motion, the result would be its mode shape.

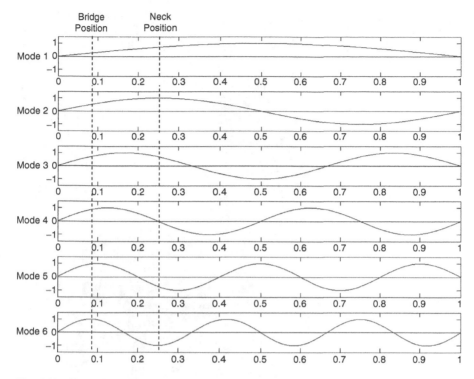

Fig. 5.22 First Six Modes of a String

Figure 5.22 shows the first 6 modes of an ideal string. Remember that the motion of a string after being plucked is a combination of all its modes. The two dashed lines on the plot correspond roughly to bridge and neck pickup positions.

It's important to remember that the pickup senses string velocity. If there is no displacement, there can't be any velocity. There is displacement for all the first six modes of the string. Also, both modes 5 and 6 are at nearly their maximum displacement. For the neck position, modes 2 and 6 are at maximum displacement (for this discussion, negative and positive displacement are equivalent), but mode 4 is at zero displacement. This means that the neck pickup won't detect mode 4 at all and this frequency will be absent from the resulting sound. The extreme case occurs if a pickup is placed at the 12th fret – the center of the string. Then, the displacements of all the even modes are zero and only the odd resonant frequencies will be present in the resulting sound.

As a practical demonstration, let's look at my modified Squier Standard Stratocaster. The original single coil bridge and middle pickups have been replaced by Seymour Duncan Classic Stacks and the neck pickup has been replaced by a Seymour Duncan Mini 59. All three are humbucking pickups, but the Mini 59 has a slightly wider aperture and is slightly less sensitive at higher frequencies. The net result is an instrument with three pickup positions and two different kinds of

Fig. 5.23 Body of a
Modified Squier Standard
Stratocaster

pickups. This is exactly the kind of tonal flexibility that electric guitars offer. Figure 5.23 shows the body of this guitar.

A simple example showing how the different pickup positions affect the tone is a first position C chord played with the pick at about the middle pickup position and recorded separately using each of the three pickups. Figure 5.24 shows the results in frequency domain

To correct for any differences in how hard the strings were strummed, the results are normalized so that the maximum value for each of the three traces is 1. It's clear that the neck pickup has the least high frequency content and the bridge pickup has the most.

When designing an electric guitar, one of the most important decisions is where to place the pickups. For the small number of instruments made with a single pickup, the decision is that much more important. For instruments intended for rock, the pickup is generally at the bridge position. Figure 5.25 shows a Fender Esquire, the precursor to the Telecaster, with a single bridge pickup. Note that the pickup is angled so that the treble side is closer to the bridge than the bass side, as on the Stratocaster and many other guitars. This tends to accentuate the treble content in the highest pitch strings and reduce it in the lowest pitch strings.

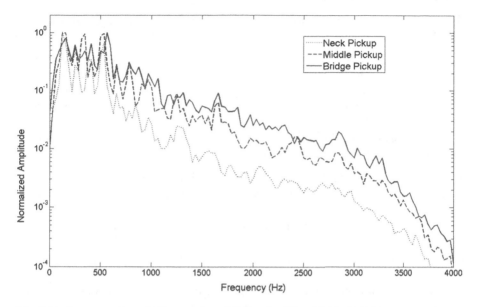

Fig. 5.24 Frequency Domain Presentation of C Chord at Three Pickup Positions

Fig. 5.25 A Fender Esquire with a Single Pickup (Wikimedia Commons, image is in the public domain)

5.3   Body Dynamics and Sound Quality in Acoustic Guitars

Perhaps the most difficult topic when discussing guitar sound quality is the tone of acoustic guitars. While the tonal characteristics of electric guitars can be controlled fairly directly by the choice of pickups and by adding electronic effects, the tone of acoustic guitars is produced by a subtle interaction of structural dynamics and acoustic radiation.

Several factors combine to make the problem of sound quality very difficult to describe or to analyze in a coherent way:

- At this writing, there is no widely accepted metric that describes sound quality of an acoustic guitar in an objective way – there is no number that measures how good a guitar sounds.
- There is no general agreement on what constitutes good sound quality since different players often value different characteristics in their instruments.
- The difference between a good acoustic guitar and an exceptional one is often very subtle.
- The connections between design changes or structural modifications and sound quality are poorly understood, or at least poorly described.

Lest this sound too discouraging, it's possible that some of the finest acoustic guitars ever made are being made now. There is a large, technically informed community of luthiers worldwide who are exploring a vast range of design options and ways to quantify sound quality. Also, factories around the world are working to both reduce the cost of new acoustic guitars and improve their quality. One happy result is that inexpensive, mass-market acoustic guitars are better than they have ever been.

At this writing, a student on a modest budget can purchase a perfectly playable instrument with a built in pickup, pre-amp and electronic tuner. The author's Fender CD-60E fits this description almost exactly and is currently available from a number of different vendors, including a nice case, for about $270.

Before venturing into the details of what we think we know about sound quality of acoustic guitars, we need to acknowledge that the relationship between musician and instrument is often not fundamentally a rational one. There are many possible examples, but one of the best may be the Martin N-20 classical guitar purchased by country singer Willie Nelson in 1969. He named it Trigger, after the horse of cowboy singer, Roy Rogers, and has played it ever since [14]. It was never designed for the heavy use that it has seen and is, by any objective measure, a complete ruin. Figure 5.26 shows the now-famous instrument with a large hole worn through the top.

Certainly, Willie Nelson can have any instrument he likes and luthiers would probably stand in line for the chance to give him one. However, none of that matters. For whatever reason, this instrument is the one for him and he shows no signs of being willing to part with it. There is no reason to think that it was ever a particularly fine instrument and Nelson apparently bought the guitar, sight unseen, as a replacement for an earlier one that was beyond repair. With this caveat firmly established, we might now proceed.

Fig. 5.26 Willie Nelson's Guitar, Trigger, a Martin N-20 Classical Guitar (Wikimedia Commons, image is in the public domain)

The most basic qualities of the ideal acoustic guitar are fairly simple to describe. It must stay in tune and produce the correct frequencies for all notes. It should radiate sound evenly throughout its register and there should be content at the higher harmonics for each note so that the resulting tone sounds full and rich. Just about any reasonably well made guitar should be able to meet these requirements. Clearly, there is more to the problem than this.

Let's divide the sound quality problem into two parts. The first one is simply descriptive – the task of describing what acoustic characteristics are correlated with subjective sound quality. The second one is to identify the design features that produce these characteristics. The descriptive portion of the problem is probably more widely represented in the academic literature, so let's start there.

5.3.1   Descriptions of Sound Quality

There is a finite amount of kinetic energy in a vibrating string that can be converted to radiated sound. When we talk about sound quality, we are essentially talking about the details of this conversion process. The vibrating string drives the top by the forces transmitted to the top through the bridge and saddle. The top's resistance to these forces is expressed as mechanical impedance (the mechanical analog to electrical impedance) and this impedance is a function of frequency. Mechanical impedance is the inverse of the transfer function.

The point of all this is that the strings drive the top and the top mostly responds proportionally to the string forces – mostly, but not completely. The structure of the guitar body has its own dynamic properties and they condition how the body responds to the forces made by the moving strings.

Fig. 5.27 Tap Testing a Classical Guitar

A good illustration of this interaction comes from tap testing a guitar. A small accelerometer is attached to some part of the guitar and it is struck with an instrumented hammer. A data acquisition system records both the hammer's input force and the accelerometer's output acceleration. Then, analysis software calculates the ratio of output to input in the frequency domain. This is the transfer function or frequency response function (FRF).

Figure 5.27 shows a tap test on a classical guitar. The instrument is supported on soft foam blocks so it is free to move however it will. In structural dynamics lingo, this is called a free boundary condition. Note that foam ear plugs have been placed between the strings. The foam used for ear plugs has extremely high damping and they very effectively attenuate the motion of the strings. If the strings were not damped, the response would be dominated by string motion and the FRF would show only string response. In order to isolate the response of the structure, the strings had to be damped.

Note that an FRF can be measured using any input location and any response location. It is certainly possible, for example, to tap the instrument at the nut and measure its response at some point on the bridge. However, the physical signifi-cance might be tough to sort out. Since the strings drive the body through the bridge, testing often focuses there. When the input and response location are the same, as shown here, the result is called a driving point FRF.

Figure 5.28 shows the FRF of the guitar with damped strings. The peaks in this curve are the resonant frequencies of the guitar body, not of the strings. When a builder chooses a design and selects materials, their effect will be a change in the

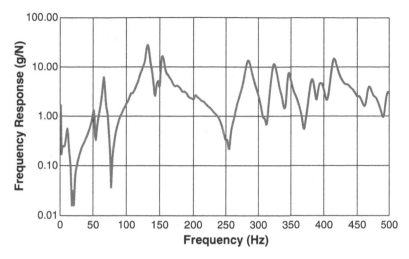

Fig. 5.28   Measured Frequency Response Function from a Classical Guitar

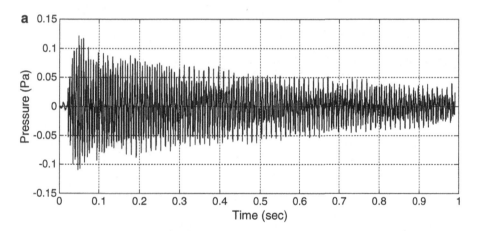

Fig. 5.29a   Response of the Low E String in Time Domain

dynamic response of structure and will be apparent in the shape of the FRF curve. One can think of the FRF as a filter that conditions the input forces from the strings and, so, colors the resulting tone of the instrument.

While describing the overall effect of complex sounds in objective terms is difficult, the characteristics of simple sounds are easier to sort out. When a string is plucked and allowed to ring, there are two different components in the resulting motion: attack and decay. Figure 5.29a shows the first second of the response of a low E string. The attack portion of the response occurs in about the first 40 milliseconds and the remainder is the decay. This data was recorded in a large hemi-anechoic chamber with very low levels of background noise.

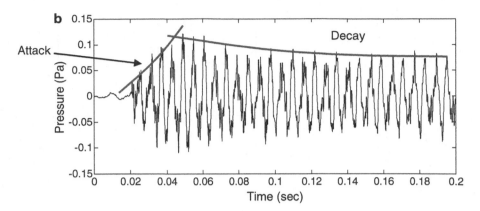

Fig. 5.29b   Attack and Beginning of Decay of Low E String

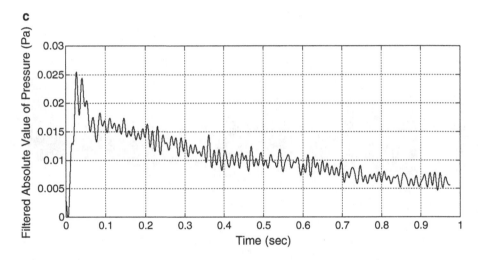

Fig. 5.29c   Low E String Data Post-Processed to Show Attack and Decay

Figure 5.29b shows the first 200 milliseconds of the signal. The signal clearly builds over the time from 0.02 seconds to 0.05 seconds, a span of 0.03 seconds (30 milliseconds). In order to make the attack and decay easier to see, the time signal can be processed further. Figure 5.29c shows the absolute value of the original time signal after the high frequency components have been removed. This is not, perhaps, a standard analysis and the resulting amplitude doesn't match that of the original signal. However, the attack and decay are clear.

Subjective evaluations have shown correlations between superior sound quality and low variation of attack times between strings [15,16]. In addition to small string to string variation in the rise time, experimental results have suggested that players prefer longer attack times – on the order of 40 milliseconds. In the pool of 13 instruments used for this testing, guitars with long attack times generally had shorter decay times.

Attack times are certainly not the only property correlated with high subjective ratings; players have also been shown to prefer high sound pressure levels (SPL) below 1 kHz. It is worth noting that the human ear is generally most sensitive around 1 kHz. High sound levels necessarily mean shorter decay times since the kinetic energy in the strings is converted to energy in the form of radiated sound; the higher the sound level, the more energy is being converted and the more rapidly the motion of the strings must decay

One important facet of the guitar sound quality discussion is whether it is being evaluated by the player or by the audience. Anecdotal evidence suggests that the player can easily distinguish between instruments while the audience can easily distinguish between players. Along these lines, Bruné [17] suggests that the sound of a guitar performance is 50% the player and 50% the guitar. It can be assumed that he was referring to classical and flamenco guitars, but I suspect his comment may apply more widely.

5.3.2   Building in Sound Quality

A very interesting list of has been proposed by Richard Bruné, a very accomplished maker of classical and flamenco guitars [17]. He lists the eight concerns of highly successful guitar makers in order of importance:

1. Environment
2. Glue
3. Model or Design
4. Strings
5. Bridge
6. Neck Setup
7. Soundboard Design
8. Varnish

What is perhaps as interesting as the contents of this list is what is omitted. He does not include materials. In fact he specifically states

> The quality of the wood is not that big an issue if you get the design elements right and you set the guitar up right. You can make a very good guitar out of mediocre wood.

To support his assertion, he offers a picture from the cover of a Christie's auction catalog which shows a violin made by Andrea Guarneri, one of the great masters, around 1680 (see Fig. 5.30). It clearly shows a knot between the f holes, near where the bridge would sit. In addition to the knot, the grain is uneven, being more widely spaced on the left than on the right. It is unlikely that any modern builder would use this piece of wood.

Fig. 5.30 A Violin Made by
Andrea Guarneri, ca 1680,
with a Knot Between the f
holes (Image courtesy of
Christie's, http://www.
christies.com)

5.3.2.1   Materials

If ever there was a discussion supporting the idea that, for every expert, there is an equal and opposite expert, the selection of wood for guitars must be it. It is clear from a volume of available literature, personal discussions and other anecdotal encounters that the selection of materials is either of critical importance, not very important at all, or something in between those two points.

It might help to start by trying to put the discussion on some kind of technical foundation. The arguments made for using only the most carefully selected materials are supported by both experiences of successful luthiers and by some basic physical properties of the materials involved. An acoustic guitar top must be both strong enough to withstand string forces that are parallel to the plane of the top and also be flexible enough to move in response to the vibration of the strings. It seems intuitive that some materials would better satisfy these opposed needs than others.

The classic choice for acoustic guitar tops is spruce, usually either Sitka spruce or Engelmann spruce. Western red cedar is also used for a minority of instruments and redwood is used more rarely. Spruce became the favored wood for guitar tops and violin tops long before mechanical properties were carefully studied, so we can

Table 5.2 Stiffness and Specific Gravity of Wood

Species – *Latin Name*	Longitudinal Modulus, E_1		Specific Gravity, SG	Stiffness/ Specific Gravity
	$\times 10^6$ psi	$\times 10^9$ Pa	unitless	GPa
Coastal Douglas Fir – *Pseudotsuga menziesii*	1.95	13.4	0.48	27.9
Sitka Spruce – *Picea sitchensis*	1.57	9.9	0.36	27.5
European Spruce – *Picea abies*	1.32	9.1	0.34	26.8
Western White Pine - *Pinus Monticola*	1.46	10.1	0.38	26.6
Engelmann Spruce – *Picea engelmannii*	1.30	8.9	0.35	25.4
Western Red Cedar – *Thuja plicata*	1.11	7.7	0.32	24.1
Red Alder – *Alnus rubra*	1.38	9.5	0.41	23.2
Redwood (Young Growth) – *Sequoia sempervirens*	1.10	7.6	0.35	21.7
American Red Spruce - *Picea rubens*	1.18	8.14	0.38	21.4
Balsa – *Ochroma pyramidale*	0.49	3.4	0.16	21.3
Sycamore – *Platanus occidentalis*	1.42	9.8	0.49	20

assume that luthiers selected it simply because they found it worked better than other species. One mechanical argument in favor of spruce is that it has a high ratio of stiffness to weight. Table 5.2 shows representative properties for several species [18]

Douglas fir actually has a higher stiffness to weight ratio than Sitka spruce, but the author has found it to be grainy, difficult to scrape or sand to a smooth surface and easily splintered – not an ideal wood for instruments in spite of its specific stiffness. Western white pine can have an even grain and is easy to work with, but is usually very soft and easily dented. In contrast, carefully selected spruce has an even grain, sands and carves easily and is hard enough to stand up to handling. It is also a nice, even light color that contrasts well with darker woods used for fretboards, backs and sides. Any builder can see why it has been so popular.

The task of selecting wood is important, though often idiosyncratic. Typically, a builder will flex a top plate half (the two halves haven't been joined yet) to get a feel for the stiffness as shown in Fig. 5.31. If the plate is judged to be stiff enough, it is then held up to the ear, held lightly between two fingers and tapped. Through this process, builders may select the top with properties they feel are best suited to the instruments they wish to make. Tops with exceptionally tight grain – closely spaced growth rings – may actually be too stiff or too 'bright' for some applications. Tops described as bright might have low damping along with high stiffness or ratio of stiffness to weight.

Explaining this process in mechanical terms is a bit speculative. It would appear that the bending test is to check the cross-grain stiffness of the wood. Since the test is done quickly, one might further assume that it is really a dynamic cross-grain test. Wood is viscoelastic, so the stiffness one measures in a bending test depends on the loading rate. A dynamic test might essentially serve simultaneously to determine stiffness and damping, at least in an approximate way.

Fig. 5.31 Flexing a Piece of Wood to Check Cross-Grain Stiffness

A sample with low damping and relatively high cross-grain stiffness will have a distinct feel when subjected to this hand testing. When tap tested, it might also sound 'bright'. A point of concern for the tap testing part of the process is that resonant frequencies of a rectangular plate are dependent on thickness of the plate. Also, the ratios of the resonant frequencies depend strongly on the aspect ratio of the plate as well. Add to this the dependence of frequencies on density, stiffness and the ratio of longitudinal to lateral stiffness and the result is a muddled picture at best.

That said, the ideal top plate has high cross-grain stiffness, low damping and a high stiffness to weight ratio. High grade spruce has even, straight grain. Additionally, grain lines should be closely spaced. Finally, the surface hardness will be relatively high. This type of wood often comes from large trees grown in a cool place, like the Pacific Northwest area of North America (Washington State, British Columbia and southern Alaska).

Processed wood and man-made materials have not yet been accepted by makers of the finest guitars, but this author is hopeful that they will. Certainly, plywood can be made from high quality wood with much less waste than occurs when quarter sawing logs. Additionally, it can be made free of knots and other imperfections that further reduce the yield of solid planks. Also, the laminations can be oriented to tailor the stiffness properties to specific needs. High quality plywood like aircraft grade plywood is very uniform, which would suit it well to series production and reduce the need to grade every piece of wood before using it. Finally, plywood is resistant to splitting so that a plywood instrument can be more durable than a solid wood one. There is every reason to think that very fine guitars can be made from laminated wood.

The case for fiber reinforced plastics is only a little harder to make. Materials like graphite, Kevlar and fiberglass have many attractive structural properties. They can be shipped and stored in rolls; material is simply cut from the roll when needed. When properly used, fiber reinforced plastics are water resistant, strong and durable. Perhaps the most difficult objection at this point is that 'plastic' guitars (usually made from carbon fiber or graphite) generally don't sound like wood guitars.

It has been my experience that composite guitars do indeed have a distinctive sound. I don't find it objectionable, though people who have spent their lives playing wood guitars can certainly be forgiven for preferring the 'wood guitar' sound. It seems likely at this writing that composite materials will probably find increasing use in guitars, though that might come only after some way is found to tailor the tone so that it more closely matches players' expectations.

An obvious solution to the problem of the increasing scarcity of top quality instrument wood is to simply use lower grade wood. Indeed, some luthiers happily use whatever wood comes conveniently to hand. There has been a rich history of instruments being made from very average wood. Certainly, folk instruments such as lap dulcimers were routinely made from whatever was available and cigar box guitars are part of the lore of folk and blues music in the US.

Well-known builders have experimented with using the most commonly available woods and some have reported making very successful instruments. Bob Benedetto made a well known archtop guitar from lumber yard wood full of knots [19] and Bob Taylor made a successful instrument of wood salvaged from a shipping pallet (Fig. 5.32). The instrument was so well received that a limited number of copies were produced for sale.

There certainly is no consensus among luthiers about what kind of materials should be used for guitars. This is perhaps a reflection of the varied tastes of both builders and of players. If so, then it is a happy commentary on the breadth of opportunities available to guitar makers. Perhaps the most concise comment on the subject of guitar wood was passed on to me by a wise luthier whose name I cannot, alas, recall:

> If you really, really know what you're doing, then you can make a good guitar from anything you like. If not, then you'd better use the best stuff you can afford.

5.3.2.2 Design Features

Since there is no broad agreement about what constitutes good sound quality or how to measure it, there should be no surprise in learning that there are many different design features intended to improve sound quality.

Broadly, there are some basic ideas that seem to have been accepted by a large number of builders, though certainly not all of them.

Fig. 5.32   The Taylor 'Pallet' Guitar (Image courtesy of Taylor Guitars, http://www.taylorguitars. com)

Since the purpose of the top is to vibrate and radiate sound, it makes sense to do what is possible, within the limits of a sound structure, to make the top free to move. One approach is to reduce the thickness near the edges of the soundboard. Taylor routs a shallow channel near the edge of the soundboard to reduce bending stiffness there. Bob Benedetto thins the edges of his arched tops to form a slight recurve near the edge, also to reduce bending stiffness. Finally, essentially all luthiers taper the top braces to nearly zero thickness at the edges of the soundboard. Figure 5.33 shows a stack of tops at the Taylor factory in El Cajon, California. They clearly show the edge routing and the tapered top braces.

Another approach to allowing the soundboard to move as freely as possible is to increase the surface area that can vibrate. There are often heavy cross bars at the waist, particularly in classical guitars. Figure 5.34 shows a classical guitar made in 2000 by Richard Bruné. The two heavy cross braces, one above the sound hole and one below, are clearly visible.

Fig. 5.33 Completed Tops at Taylor Guitars (Image by the author, reproduced here courtesy of Taylor Guitars, http://www.taylorguitars.com)

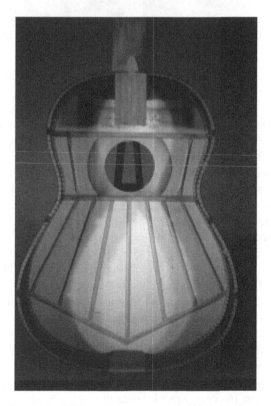

Fig. 5.34 Bracing of a Classical Guitar (Image courtesy of Richard Bruné, http://www.rebrune. com)

a

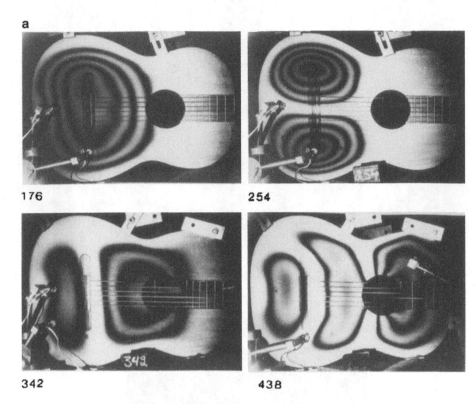

176   254

342   438

Fig. 5.35a First Four Modes of a Classical Guitar (Image courtesy of Karl Stetson, http://www.holofringe.com)

These braces greatly limit the motion of the top in the upper bout. This is clearly apparent in a series of time averaged holograms by Karl Stetson [20] shown in Fig. 5.35a and 5.35b. The dark fringes are lines of constant displacement. The resonant frequencies appear below each image and each image is the experimentally determined mode shape corresponding to that frequency. The bracing pattern is unknown, but is likely to be fan bracing. The mode shapes are symmetric, suggesting that the bracing pattern is as well. There is only significant upper bout motion in the fourth mode.

On classical guitars, the cross braces are needed to keep the body from buckling at the waist. Even the small 1863 Torres in Fig. 1.16, a guitar with no soundboard bracing at all, still has cross braces. The challenge, then, is how to preserve the stiffness of the cross braces while not overly constraining top motion. One approach is to undercut the top braces so that they contact the top only in the center and edges. Figure 5.36 shows the open cross braces on a classical guitar made by Jeff Elliott. All three are significantly carved away to allow motion in the upper bout. Some builders even use arched braces sometimes called flying braces or flying buttress braces. To prevent splitting, there are 0.5 mm (0.0197 in) thick reinforcements across the grain where the cross braces would otherwise be glued on. He reports no cracking there after having used this design for more than 21 years.

b

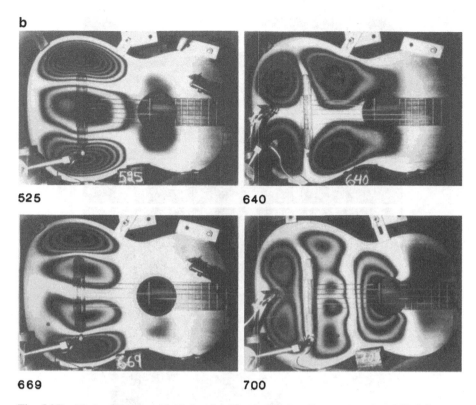

525

640

669

700

Fig. 5.35b Modes Five through Eight of a Classical Guitar (Image courtesy of Karl Stetson, http://www.holofringe.com)

Elliott also notes that this design seems to mature faster than more conventional designs. It is generally accepted that acoustic guitar tone changes as the instrument ages [21]. Acoustic guitars are generally perceived to have better tone after they have been played for some time. The effect is a subtle one, but generally accepted to be a real one. One often hears a player or builder saying that an instrument has 'opened up' after extended playing. This is a frustratingly vague term but seems to indicate that the frequency content of the instrument has changed over time. The body of technical literature on the subject is still small and deals mostly with violins [22]. It should be noted that not all experiments found a clear difference in instruments as a result of having been played [23].

This is a particularly innovative and productive time for luthiers; not only are some outstanding builders now practicing, but many are taking the time to write detailed descriptions of what they have learned. Some of these books are as beautiful as they are informative. One theme that emerges is that details can be important. Let's review some of these details.

One design detail is ensuring a solid mechanical connection between the strings and their end supports – the nut and saddle. The break angle over the nut should be

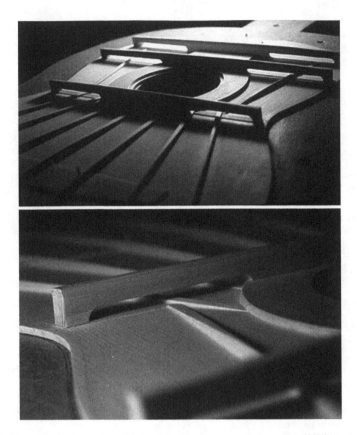

Fig. 5.36 Open Cross Braces in a Classical Guitar (Images courtesy of Jeff Elliott, http://www. jeffelliott.com)

large enough to firmly seat the string. A break angle on the order of 15° - 20° is typical, as shown in Fig. 5.37. Significantly less can give poor contact with the nut, perhaps allowing the string to buzz. Significantly more can cause tuning problems as the strings cannot move smoothly through the nut slots.

As shown in Fig. 5.37, the break angle can depend on the design of the instrument. For a guitar with a solid headstock, the angles for all the strings can be approximately the same. However, if the headstock is slotted, the angle is different depending on which tuner the string is fixed to.

On the other end of the string, it is equally important to create a sharp break angle at the saddle. Here, there is no requirement for the string to slide smoothly through a slot while being tuned. In classical guitars, some builders use two holes in the tie block for each string. This allows the string to angle directly toward the tie block hole without being raised by the traditional loop and knot [24]. Figure 5.38 shows the bridge on a classical guitar made by Jeff Elliott that uses the double hole method of anchoring the strings.

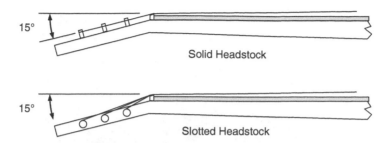

Solid Headstock

Slotted Headstock

Fig. 5.37 Break angle over the nut

Fig. 5.38 A Classical Guitar Bridge with Two Sets of Holes in the Tie Block (Image by the author, reproduced here courtesy of Jeff Elliott, http://www.jeffelliott.com)

Another way to increase the break angle at the bridge is to raise the saddle. However, this would increase the moment (torque) acting on the bridge and, thus, the top. In the extreme case, this can cause a crack in the top.

On a steel string guitar, slots can be cut in front of the bridge pin holes to increase the break angle over the bridge. Figure 5.39 shows such a slot cut into a factory-made bridge. The other five slots were cut after this picture was taken.

Finally, another suggestion by Ervin Somogyi, that he believes originated with John Gilbert, is that the saddle should be rounded to allow more contact area with the string. This might require a wider saddle than is typical (Fig. 5.40).

From smaller details, let's progress to the larger ones. Since the body transforms string vibrations into radiated sound and is the heart of an acoustic guitar, it is the component that has attracted the most attention from luthiers seeking to improve sound quality. Perhaps surprisingly, one part of the body that has attracted interest is the sides.

Several makers report that heavy or stiff sides improve the sound of steel string acoustic guitars. The R. Taylor line of acoustic guitars, the top line made by Taylor Guitars, uses heavy, bent kerfing and additional stiffeners for the sides, as shown in Fig. 5.41. The R. Taylor side in the foreground is compared to a standard production Taylor side with conventional slotted kerfing.

Fig. 5.39 A Single String Slot Cut by the Author into a Steel String Acoustic Bridge

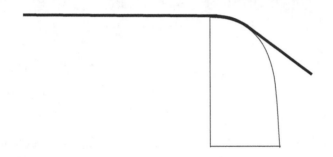

Fig. 5.40 A Rounded Saddle that Increases String Contact Area (after Somogyi)

Along these lines, Trevor Gore [25] has experimented with adding masses to the sides to greatly increase their resistance to motion. Energy not being used to move the sides, which are poor acoustic radiators, can instead move the top, which is a good radiator. The instrument I saw a few months before this book went to press had at least one steel block fixed to the inside of the lower bout. I thought the sound quality was quite good, though the weight of the instrument reflected the fact that the sides had been ballasted.

The last and probably most controversial topic in the sound quality of acoustic guitars is that of tuning or voicing the body. Some builders devote significant effort to modifying the structure of their instruments so that they conform to an ideal of tone, dynamic response or frequency ratios. In many cases, the assumed physical principles are expressed in idiomatic terms, difficult to transform into a rigorous mathematical discussion. Even so, luthiers working with personally developed processes for voicing their instruments routinely produce very good guitars.

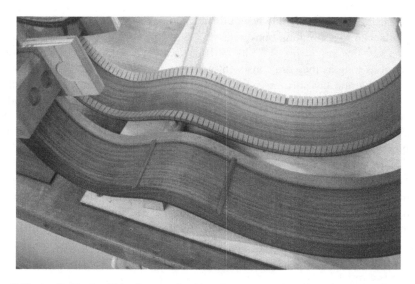

Fig. 5.41 An R. Taylor Side Compared with a Standard Taylor Side (Image by the author, reproduced here courtesy of Taylor Guitars, http://www.taylorguitars.com)

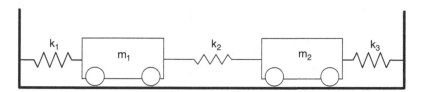

Fig. 5.42 Two Coupled Oscillators

The range of ideas on voicing instruments is far beyond this discussion, so let's stick to a manageable number of basic ideas and to ones that can be expressed unambiguously. There seems to be general agreement that a luthier trying to tune the frequency response of an acoustic guitar body can't reasonably expect to explicitly control more than the first few frequencies. Efforts to voice guitar bodies typically focus on these lower modes.

The first resonant frequency of full-sized acoustic guitars is generally in the range of 95 Hz – 105 Hz and the second frequency is generally near 200 Hz. In addition to this broad guideline, more specific recommendations have been suggested. A clear suggestion [22] is that none of the body resonant frequencies should lie close to a note frequency. It is well-known that coupling two vibrating masses can change their frequencies, even when that coupling is weak. Figure 5.42 shows the kind of spring-mass system that students work with when learning structural dynamics. The carts just represent masses that can move horizontally without friction. The end springs, k_1 and k_3, are much stiffer than the coupling spring, k_2. The carts represent, in a simple way, the mass of the top and the mass of a moving string. In a very simple way, this mechanical system represents a vibrating string attached to a vibrating body.

Let's say that $m_1 = m_2 = 1$, $k_1 = 1000$ and $k_3 = 1000.5$. The units aren't important as long as they are consistent with one another. If you like, assume the mass is in kg and the stiffness is in N/m. If the stiffness of the coupling spring is zero, then the two masses vibrate completely independently of one another. The resonant frequency of the first cart is 31.62 rad/sec and the resonant frequency of the second cart is 31.63 rad/sec. This is analogous to a body resonant frequency that closely corresponds to a note frequency.

Now, let's assume that the stiffness of the coupling spring is increased so that $k_2 = 10$. It is a factor of 100 smaller than the other two, so this is analogous to the weak mechanical coupling between the strings and body of a guitar. In this case, the first resonant frequency of the system is 31.63 rad/sec and the second resonant frequency of the system is 31.94 rad/sec. The first frequency changed very little, but the second one increased by about 1%. This is the kind of coupling that could make a note sound at the wrong frequency, even though the string is in tune and the fret spacing is correct.

Say, now, that we change the stiffness so that $k_1 = 900$ and $k_3 = 1100$. Without any coupling, the first resonant frequency is 30 rad/sec and the second one is 33.17 rad/sec. Now, the frequencies are separated by 10.6%. With the same coupling as used before, the new resonant frequencies are 30.16 rad/sec and 33.32 rad/sec. The maximum change in frequency now is a little less than 0.5%.

A related design suggestion [22] is that the first two body modes should not be exactly an octave apart – the second resonant frequency should not be exactly twice the first. Otherwise, there could be a single note that sounds much more loudly than the others. Remember that a good guitar will not have any particularly loud or soft notes.

Among the many proposed criteria for voicing acoustic guitars, perhaps the most clearly expressed is a relatively recent one by Trevor Gore, that of monopole mobility. The mode shapes of guitar tops are sometimes described in terms of how many radiating areas they have. In the holograms from Fig. 5.35a, the first mode at 176 Hz (which may actually be the second mode of the body) has a single portion of the body in motion. Thus this mode forms an acoustic monopole.

The next mode at 254 Hz has a node line down the center. While it is not obvious from this type of image, the two halves of the soundboard are moving in opposite directions. In acoustic lingo, they are out of phase with one another. Thus, there are two clearly defined portions of the top that are acoustically radiating and this mode is, thus, an acoustic dipole.

The second bit of terminology to sort out is mobility. This has a very specific technical definition that is different from the popular definition you might find in a dictionary. Rather, it has to do with how frequency response functions (FRFs) are defined. An FRF defined as acceleration per unit force is traditionally called accelerance. This is the most common form of FRF in structural dynamics since response is usually measured with accelerometers. An FRF defined as velocity per unit force is called mobility. This is of particular interest in structural acoustics problems (like guitar design) since the radiated sound is determined by the velocity of the moving surface. Just for completeness, the FRF defined as position per unit force is called receptance.

Gore posits that high monopole mobility – a measure of how readily the vibrating strings can induce velocity in the two lowest frequency modes of the top – is an indicator of a good guitar [25]. However, high monopole mobility is not enough by itself. The instrument must also:

- Play in tune across all strings and frets
- Be even in volume across the strings and frets
- Have an alluring tone

5.4   Diversity of Designs

It does seem sometimes that, for every expert luthier, there is an equal and opposite expert. So how is it that different luthiers can suggest different and sometimes opposing approaches to making fine instruments? I can only offer a hypothesis drawn from experience with design optimization.

When solving an optimization problem, there are design variables that can be altered at will within the limits of physical constraints, like maximum allowable stress. For example, if the object is to design a wind turbine that minimizes the cost of the electricity produced, the designer can select the number of blades, their geometry, the height of the hub above the ground and many other characteristics. The quantity to be minimized – like cost of electricity - is called the objective function.

In mathematical terms, each design variable is considered as a separate, orthogonal dimension in space. The result is called design space and it has as many dimensions as there are design variables. Finding an optimal solution for a problem that has, say, 25 design variables means finding the point in 25 dimensional design space where the objective function (cost of electricity from the wind turbine) is the lowest.

For mathematical reasons, optimization problems always look for the minimum of the objective function; if you need to maximize some value (like sound quality), you simply minimize its negative. In almost all meaningful optimization problems, there are many points in design space where the objective function is close to the minimum. This means that there are usually many points, often widely separated, that would yield a nearly optimal design. The designers generally need only find one of these points, called local minima. Figure 5.43 shows a hypothetical design space with two design variables. There are several local minima.

A luthier may have dozens of design decisions to make, and every one of them represents another dimension of design space. It is quite reasonable to assume that there are multiple points in that design space that would give good sound quality and they might be widely separated (i.e. be very different designs). If so, then that luthier need only find one of those points.

This hypothesis definitely falls into the category of things I believe, but can't prove. However, it is compatible with design results in many other areas. It is the reason that all cars don't look alike, all chairs don't look alike and so on. If this hypothesis is

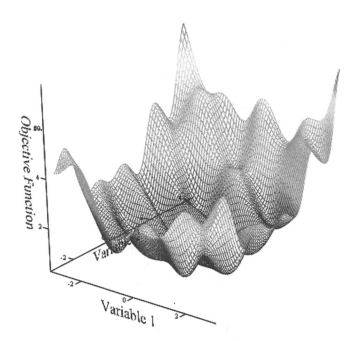

Fig. 5.43 A Two Variable Design Space with Several Local Minima

correct, then there are many, mutually exclusive design decisions that could result in a fine guitar. In more practical terms, two builders making opposing design decisions might both produce fine guitars – they can both still be approximately correct.

As an example, let's consider the wood chosen for instruments. As discussed earlier, builders generally choose wood carefully. Consciously choosing materials with different properties can distinctly change the tone of the instrument and could well be considered a design change. Of course, it helps that different players can have very different opinions on what constitutes a good guitar.

At C.F. Martin, to highlight one of the oldest and most successful manufacturers, there is a well-developed awareness of the differences in instruments of identical design, but different weights. A particularly light one was described as having a "powerful, breathy, crystalline, glassine clarity" that made it a wonderful instrument. Conversely, a heavier instrument can have resonance and warmth that makes it equally, though differently, wonderful [26].

References

1. Bartolini W and Bartolini PA (1982) Experimental Studies of the Acoustics of Classic and Flamenco Guitars. Journal of Guitar Acoustics, 6
2. Gilbert J (1984) Intonation and Fret Placement. Soundboard 26–27
3. Byers G (2006) Classical Guitar Intonation, in Big Red Book of American Lutherie, Vol. 4 1994–1996. Guild of American Luthiers

4. Varieschi GU and Gower CM (2010) Intonation and Compensation of Fretted String Instruments. American Journal of Physics 78(1)
5. Elmendorp S (2010) It's All About the Core – Or How to Estimate Compensation. American Lutherie 104 56–60
6. Gore T (2011) Contemporary Guitar Design and Build – Volume 1: Design
7. Magliari G and MacRostie D (2011) Beyond the Rule of 18: Intonation for the 21st Century Guild of American Luthiers 2011 Convention, Tacoma
8. D'Addario Web Site, http://www.daddario.com, last visited Aug 5, 2011
9. Earvana Web Site, http://www.earvana.com, last visited Aug 1, 2011
10. Buzz Feiten Web Site, http://www.buzzfeiten.com, last visited Aug 1, 2011
11. Personal conversation, May 2011
12. True Temperament Web Site, http://www.truetemperament.com, site last visited Aug 5, 2011
13. Lehman B (2005) Bach's Extraordinary Temperament: Our Rosetta Stone. Early Music, 33(1), pp 3–23
14. Patoski JN (2008) Willie Nelson: An Epic Life. Little, Brown and Company
15. Jaroszewski A, Rakowsky A and Sera J (1978) Opening Transients and the Quality of Classical Guitars. Archives of Acoustics 3(2) 79–84
16. Orduña-Bustamante F (1992) Experiments on the Relation between Acoustical Properties and the Subjective Quality of Classical Guitars. Journal of the Catgut Acoustical Society 2(1) 20–23
17. Bruné RE (2004) Eight Concerns of Highly Successful Luthiers. American Lutherie 79 6–21
18. wood handbook, Green DW Winandy JE and Kretschmann DE, Wood Handbook – Wood as an engineering Material, Forest Products Lab 1999 FPL-GTR-113 Forest Products Lab
19. Benedetto R (1994) making an Archtop Guitar. Centerstream Publishing
20. Stetson KA (1981) On Modal Coupling in String Instrument Bodies. Journal of Guitar Acoustics, 3: 23–31
21. Johnson R (2010) Why Do Guitars Sound better as They Age? Acoustic Guitar
22. Hutchins CM (1998) A Measurable Effect of Long-Term Playing on Violin Family Instruments. Catgut Acoustical Society Journal, 3(5) pp 38–40
23. Inta R, Smith J and Wolfe J (2005) Measurement of the Effect on Violins of Aging and Playing. Acoustics Australia 33(1)
24. Somogyi E (2009) The Responsive Guitar. Luthiers Press
25. Gore T (2011) Wood for Guitars. Proceedings, 161st Meeting of the Acoustical Society of America
26. Dick Boak, private correspondence

Chapter 6
Design Specifics for Acoustic Guitars

There are many different acoustic guitars now available. While there are far too many to cover completely here, most of them can be grouped into a few categories. In the following sections, representative descriptions of the instruments are listed along with design features and dimensions.

6.1 Classical Guitars

Classical guitars usually follow traditional designs and, to a casual viewer, can look surprisingly uniform. Richard Bruné, a maker and restorer of fine classical guitars, suggests the proportions for the body shown in Fig. 6.1 [1].

The dimensions are all relative to the body length, which is normalized to 1. A typical body length is 490 mm and the rest of the dimensions are proportional. For example, a 490 mm body length would yield a waist width of 245 mm (0.5 × 490 mm) and an upper bout width of 279 mm (0.57 × 490 mm). Table 6.1 gives a list of dimensions for a representative classical guitar.

One dimension not listed in Table 6.1 is the amount by which the nut end of the neck is raised above the plane of the top. This distance varies from builder to builder, but is often on the order of a few millimeters as shown in Fig. 6.2.

It is helpful to have pictures to show the details of a well-made, traditional classical guitar. Figure 6.3 shows the body of a nice classical guitar made by Jose Bellido in Granada, Spain and now owned by Prof. Eugene Coyle at Dublin Institute of Technology. This instrument also appears in Figs. 6.5, 6.6, 6.7 and 6.9.

Figure 6.4 shows the back of this instrument. It is made from book matched pieces of rosewood and has the traditional decorative center strip. The contrasting colors of the multi-part wood binding are also clearly visible.

Figure 6.5 shows a close up view of the heel and upper body of the guitar. This instrument was made in the traditional manner with the neck and body being an integral unit (as opposed to being made separately and then joined). Note how the

R.M. French, *Technology of the Guitar*, DOI 10.1007/978-1-4614-1921-1_6,
© Springer Science+Business Media New York 2012

Fig. 6.1 Suggested
Proportions for a Classical
Guitar Body

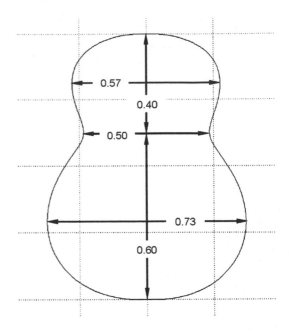

Table 6.1 Representative Dimensions for a Classical Guitar

Dimension	Value – inches	Value – mm
Scale Length	25.5	648
Headstock Width – Top	2.89	73.4
Headstock Width - Bottom	2.47	62.7
Headstock Thickness	0.875	22.2
Nut Width	2.125	54.0
Fretboard Thickness	0.25	6.35
Neck Width at 12th Fret	2.71	68.7
Neck Thickness at Nut (excluding the fretboard thickness)	0.688	17.5
Body Length	19.25	489
Soundhole Diameter	3.5	88.9
Soundhole Center Location	18.7 from nut 5.93 from 12th fret	475 from nut 151 from 12th fret
Body Depth @ Tail Block	3.75	95.3
Body Depth @ Neck Block	3.5	88.9
Upper Bout Width	11	279
Width at Waist	9.65	245
Lower Bout Width	14.1	357
Nominal Soundboard Thickness	0.100	2.5
Side Thickness	0.080	2.0
Back Thickness	0.095	2.4
Tuner Spacing	1.378	35
String Spacing at Nut	1.81	46.0
String Spacing at Saddle	2.36	59.9

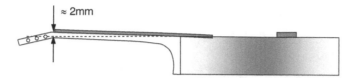

Fig. 6.2   Distance of Neck above the Plane of the Top on a Classical Guitar

Fig. 6.3   Front of Body of a Representative Classical Guitar

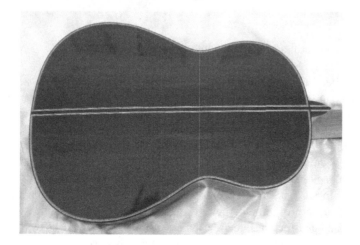

Fig. 6.4   Back of Body of a Representative Classical Guitar

Fig. 6.5 Heel Showing Stacked Construction

Fig. 6.6 Soundhole and Rosette Showing Builder's Label

binding has been carefully fitted right up to the neck and heel. This is potentially difficult since this instrument was made in the traditional way with the internal foot and the sides let directly into the heel structure. Note also the multiple pieces glued together to form the heel. The neck and heel blank were clearly made from a plank around 20 mm (a little more than ¾ in) thick.

Figure 6.6 shows a close up view of the rosette and the builder's label on the inside of the back plate. Classical rosettes like this one are traditionally made from very small colored sticks of wood glued together to form an intricate pattern. Since classical guitars are often superficially quite similar, details such as the rosette design and the headstock design are used essentially as trademarks by the builders. It is very bad form indeed to copy a builder's headstock or rosette without permission.

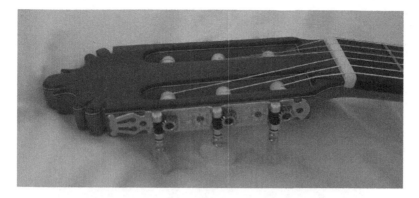

Fig. 6.7 Headstock Showing Laminations and Decorative Carving

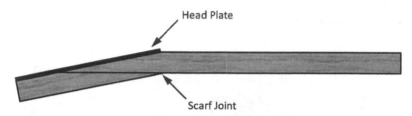

Fig. 6.8 Traditional Headstock Joint on a Classical Guitar

Figure 6.7 shows the headstock with its decorative carving at the top. The mahogany is covered by a laminated plate with several thin layers behind a thicker rosewood top plate.

In addition to aesthetics, the headstock plate serves an important structural purpose. The headstock angle is created by a scarf joint between the headstock and the neck as shown in Fig. 6.8. Remember that glue is most effective in shear. The tension of the strings on the roller exerts a shear stress on the headstock joint. The cover plate adds significantly more gluing area and further strengthens the joint.

Finally, Fig. 6.9 shows the bridge with the saddle and tie block. The builder has added a small piece of clear plastic, presumably of the material often used for pick guards, behind the bridge to keep the string ends from scratching the top.

6.2   Steel String Guitars

If classical guitars are relatively uniform, steel stringed guitars are almost the opposite. These instruments have been made in a wide range of sizes and shapes. Fortunately, there are a few recognizable categories that can accommodate most instruments.

Fig. 6.9 Bridge Showing Tied On Strings

Fig. 6.10 A Martin 00-30 Dating from 1906-7

6.2.1 Parlor Guitars

The term parlor guitar stems from a type of small-bodied guitar intended to be played at home (in the parlor). Perhaps the most familiar instruments in this category are the Martin Size 1, Size 2, Size 0, and perhaps Size 00 style guitars popularized between 1860 and 1930. Figure 6.10 shows a Martin 00-30 dating

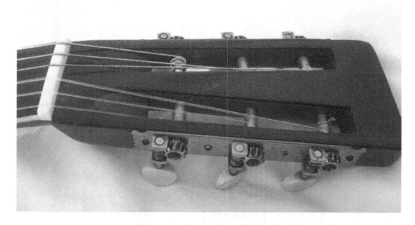

Fig. 6.11 Headstock of Martin 00-30

from 1906-7 (serial number 10686). It is currently owned by Prof. Eugene Coyle of Dublin Institute of Technology. Next to the guitar is a rare hard leather case that appears to be original. This instrument was designed to use steel strings, but has been fitted with nylon strings to lower string tension.

Figure 6.11 shows the headstock of the Martin 00-30. This instrument has a slotted headstock as steel stringed instruments often did at the time. They are not as commonly used today, with most manufacturers preferring solid headstocks with independent tuners. More representative of modern instruments is the angled headstock. There is no standard angle, but almost all acoustic guitars and some electric guitars have the headstock angled 13° - 15° from the plane of the fretboard. Note also the extremely simple Martin headstock; this design is still used on Martin guitars.

Figure 6.12 shows a close up of the body of the Martin 00-30. The sound hole is decorated with dark wood rings surrounding a mother of pearl inlay. While it is hard to tell from the inlays on the fretboard, the neck on this instrument crosses the body at the 12th fret.

Finally, Fig. 6.13 shows the heel and back of the body. The neck and heel are made of mahogany and the body is rosewood. Note the heel cap and the decorative center stripe in the back.

As this is being written, small guitars are becoming more popular and several newer designs are being successfully marketed. Taylor has been successful with the GS-Mini, a small instrument modeled on their larger instruments (Fig. 6.14). They have also produced a limited edition parlor guitar (Fig. 6.15) made along more traditional lines, including a 12 fret neck.

Finally, Martin Guitars makes their LXM series of instruments, modeled after the 0-14 instruments (Fig. 6.16).

Table 6.2 presents dimensions of these small guitars. Their smaller dimensions make them more portable and also more suitable for smaller players, including children.

Fig. 6.12 Body of Martin 00-30 Showing Soundhole

Fig. 6.13 Back of Martin 00-30 body

6.2.2   *Full-Sized Guitars*

The term 'full-sized guitars' is a catch-all description of a range of acoustic guitars that are not Dreadnoughts and have a scale length in the neighborhood of 25 in (635 mm). The archetype for these instruments might be considered as the Martin OM series discussed earlier. Figure 6.17 shows a Taylor GA (Grand Auditorium) model guitar that resembles the Martin OM.

Fig. 6.14 Taylor GS Mini (Image courtesy Taylor Guitars, http://www.taylorguitars.com)

Fig. 6.15 Taylor 35th Anniversary Parlor Guitar (Image courtesy Taylor Guitars, http://www.taylorguitars.com)

Fig. 6.16 Martin LXM Guitar (Image courtesy C.F. Martin Archives, http://www.martinguitar.com)

Table 6.2 Dimensions of Parlor-Sized Guitars

	Martin 00 – 14 Fret		Taylor GS-Mini		Taylor Parlor		Martin LXM	
	Inches	mm	Inches	mm	Inches	mm	Inches	mm
Overall Length	38.63	981	36.63	930	38.38	975	34	864
Scale Length	24.9	632	23.5	597	24.88	632	23	584
Body Length	18.88	479	17.63	448	19.5	495	15	381
Lower Bout Width	14.31	364	14.38	365	14.1	358	12	309
Nut Width	1.75	44.5	1.69	42.9	1.75	44.5	1.69	42.9
String Spacing (Bridge)	2.25	57.2					2.13	54.0
Body Depth	4.13	105	4.44	113	4.2	107	3	76.2

Fig. 6.17 Taylor GA Model Acoustic Guitar (Image courtesy of Taylor Guitars, http://www. taylorguitars.com)

It is very common for acoustic guitars to have a cutaway that allows better access to the higher frets. There are two types of cutaways commonly using in acoustic guitars, Florentine and Venetian. The names seem to have been marketing terms and probably don't refer to designs from Florence or Venice. The Venetian cutaway is rounded, like the one on the Taylor 414CE in Fig. 6.18. The body side with the cutaway is bent from a single piece of wood.

The Florentine cutaway comes to a point and the resulting side is made from two pieces of wood. Figure 6.19 shows a Washburn EA10SDLK with a Florentine body cutaway and onboard electronics.

Finally, it helps to have a list of basic dimensions for representative full-sized instruments. Table 6.3 lists basic dimensions for three different designs. Note that the Taylor GC (Grand Concert) body is similar, but a bit smaller than the Taylor GA (Grand Auditorium). Also, the dimensions for the Martin OM are for modern production models and may differ slightly from older versions.

Fig. 6.18 Taylor 414CE with Body Venetian Cutaway (Image courtesy of Taylor Guitars, http://www.taylorguitars.com)

Fig. 6.19 Washburn EA10SDLK with Florentine Cutaway (Image courtesy of Washburn Guitars, http://www.washburn.com)

Table 6.3 Dimensions of Full-Sized Guitars

	Martin OM (Modern)		Taylor GA		Taylor GC	
	Inches	mm	Inches	mm	Inches	mm
Overall Length	39.81	1011	41	1041	40.5	1029
Scale Length	25.4	645	25.5	648	24.88	632
Body Length	19.38	492	20	508	19.5	495
Lower Bout Width	15	381	16	406	15	381
Nut Width	1.75	44.5	1.75	44.5	1.75	44.5
String Spacing (Bridge)	2.25	57.2				
Body Depth	4.13	105	4.63	117	4.38	111

Fig. 6.20 Martin HD-28
Dreadnought Guitar (Image
courtesy of C.F. Martin
Archives, http://www.
martinguitar.com)

6.2.3 Dreadnought Guitars

It's hard to spend much time looking at acoustic guitars without coming across dreadnoughts. At this writing, they are very popular and may even be the majority of acoustic guitars now being sold. Figure 6.20 shows a Martin HD-28, an archetype Dreadnought guitar.

Many Dreadnoughts include cutaways and on-board electronics design. Figure 6.21 shows a Washburn WD55SCE with both electronics and a cutaway. The traditional square-shouldered design is obvious. Indeed, it is very similar to other dreadnoughts, with the primary differences being the details that identify it as a Washburn. The headstock on this instrument is one of only a few designs used by Washburn on their acoustic guitars; another is shown on the instrument in Fig. 6.19. Compare it to the headstock design of the Taylor 414CE above; it uses the same headstock design seen in essentially all Taylor acoustics.

Table 6.4 lists the basic dimensions of two common dreadnought guitars.

In addition to dimensions for specific guitars, it is helpful to have dimensions for a notional full-sized guitar. Table 6.5 presents nominal dimensions for such an instrument. Specific instruments can vary from these dimensions, but they are a good starting point.

Fig. 6.21 A Washburn WD55SCE Dreadnought Acoustic Guitar (Image courtesy of Washburn Guitars, http://www.washburnguitar.com)

Table 6.4 Dimensions of Dreadnought Guitars

	Taylor 410CE		Martin D-28	
	Inches	mm	Inches	mm
Overall Length	41	1041	40.5	1029
Scale Length	25.5	648	25.4	645
Body Length	20	508	20	508
Lower Bout Width	16	406	15.63	397
Nut Width	1.75	44.5	1.69	42.9
String Spacing (Bridge)			2.13	54.0
Body Depth	4.63	117	4.88	124

Finally, any guitar needs to have the correct setup in order to play well. Table 6.6 presents typical string heights and relief for a full-sized acoustic guitar. Again, these numbers are only meant to be representative, but are a good starting point.

6.2.4   Jumbo Guitars

The largest common size for acoustic guitars is the jumbo body. These instruments are less popular than other types, presumably because they are more difficult for small players to manage and more difficult to transport. Because of their size, they can be louder than smaller guitars – not much of an advantage if the instrument is amplified – and can have a deeper tone.

The pattern for this type of guitar was established in the late 1930s with the introduction of the Gibson Super Jumbo, now designated the J-200. Certainly, Gibson and Epiphone are not the only manufacturers to produce jumbo guitars. Many product lines include jumbo instruments, including Washburn and Taylor. Figure 6.22 shows a Washburn WJ45S.

Table 6.5 Typical Dimensions of a Dreadnought Acoustic Guitar

Dimension	Value – inches	Value – mm
Headstock Width – Top	3.875	98.4
Headstock Width - Bottom	3.25	82.6
Headstock Thickness	0.600"	15.2
Nut Width	1.75	44.5
Fretboard Thickness at Center	0.250	6.35
Fretboard Radius	16	406
Neck Width at 12th Fret	2.25	57.2
Neck Thickness at First Fret (including the fretboard thickness)	0.850	21.6
Body Length	20.0	508
Soundhole Diameter	4.00	102
Soundhole Center Location From Top of Body	5.93	151
Body Depth @ Tail Block	4.40	112
Body Depth @ Neck Block	3.55	90.2
Upper Bout Width	11.75	298
Width at Waist	10.82	275
Lower Bout Width	16	406
Nominal Soundboard Thickness	0.115	2.92
Side Thickness	0.100	2.54
Back Thickness	0.100	2.54
Tuner Spacing	1.63	41.3

Table 6.6 String Heights for a Steel String Acoustic Guitar

	String 6 – Bass E		String 1 – Treble E	
	Inches	mm	Inches	mm
1st Fret	0.023	0.58	0.013	0.33
12th Fret	0.090	2.29	0.070	1.78
Relief at the 8th fret	0.002	0.05	0.002	0.05

Fig. 6.22 Washburn J45S Jumbo Guitar (Image courtesy of Washburn Guitars, http://www. washburnguitar.com)

Fig. 6.23 A Taylor 815CE Jumbo Guitar (Image courtesy of Taylor Guitars, http://www. taylorguitars.com)

Table 6.7 Dimensions of Taylor 615E Jumbo Guitar		Taylor 615CE	
		Inches	mm
	Overall Length	42	1067
	Scale Length	25.5	648
	Body Length	21	533
	Lower Bout Width	17	432
	Nut Width	1.75	44.5
	Body Depth	4.63	117

Figure 6.23 shows a Taylor 815CE with on-board electronics and a Florentine cutaway. Table 6.7 presents dimensions for this instrument.

6.3 Resonator Guitars

A specialized type of acoustic guitar is the resonator guitar. In this type, the flexible soundboard is replaced with one or more metal cones that vibrate in response to the string motion. There have been several different resonator designs, but probably the most familiar one was originally designed by John Dopyera and manufactured by Dobro Manufacturing Company, a company he founded with his brothers. Dobro was a contraction of Dopyera Brothers and, conveniently, also meant 'goodness' in their native Slovak language [2]. The other major manufacturer was National Reso-Phonic, which still makes resonator guitars of basically two different types: a single resonator and triple resonator called a tricone. Gibson now owns the Dobro name.

Many manufacturers make resonator guitars, usually based on the Dobro style. Figure 6.24 shows a model R15RCE single resonator guitar by Washburn. The overall shape is conventional, though the body is dominated by the resonator cover.

Fig. 6.24 A Model R15RCE Resonator Guitar by Washburn (Image courtesy of Washburn Guitars, http://www.washburn.com)

Fig. 6.25 A National Tricone Resonator Guitar (Image by the author, reproduced here courtesy of Dave Baas, http://www.roadworthyguitars.com)

Note the 12 fret neck and the single f-hole. In guitars without a cutaway, there is usually a second f-hole. It is also common for resonator guitars to have two small soundholes fitted with metal screens. Note that this instrument has a magnetic pickup between the neck and the resonator cover.

Figure 6.25 shows a National Tricone resonator guitar. The body is aluminum and the three cones are just visible behind the screens surrounding the bridge.

The dominant feature of a resonator guitar is the flexible metal cone that produces most of the sound. Figure 6.26 shows the spun metal cone and the cast 'spider' that both holds the bridge and mechanically connects the bridge to the cone.

Fig. 6.26 Resonator and Spider Visible with the Cover Removed (Wikimedia Commons, image is in the public domain)

Fig. 6.27 Mechanical Connection between the Spider and Cone (Wikipedia Commons, image is in the public domain)

Figure 6.27 shows the mechanical connection between the relatively stiff spider and the flexible cone.

While it is certainly possible to play many different styles on any instrument, including a resonator guitar, these instruments are most closely associated with blues and bluegrass styles. In these roles, they are often played with metal or glass slides.

Additionally, they are often played laying flat with soundboard facing up. Some instruments even have a rectangular neck cross-section section since a rounded cross-section is not necessary when playing with a slide and the guitar laid flat.

6.4   Archtop Guitars

Archtop guitars have been popular since the early 1900s and are used for a wide range of music now, though they are not nearly as popular as flat top acoustic guitars. The most important distinction for archtops is whether they have hollow bodies or not.

Many instruments that have arched tops and backs and f-holes are not really acoustic instruments because they have solid blocks joining the tops and backs. These blocks typically run down the centerline of the instrument. They serve as mounting structures for the pickups and bridge and also limit the possibility of feedback. However, a center block also greatly limits the motion of the soundboard and means that the instrument is only practical when amplified. For an archtop to be truly an acoustic instrument, it must have a hollow body. Electric archtops with center blocks will be covered in a later chapter.

Figure 6.28 shows an Au Naturel model acoustic archtop by Linda Manzer. The scale length of this instrument can vary slightly based on customer preferences and ranges from 64 cm – 65 cm (25.2 in – 25.6 in). The body width is 17.25 in (438 mm). While larger instruments have been made, this is near the upper size limit for most players.

This instrument shows many of the characteristic features of archtop guitars. It has two f-holes in place of the round central sound hole usually found on flat top instruments. It also has a floating bridge with a tailpiece. A floating bridge is one held in place only by the downward component of the string tension. If all the strings are removed, a floating bridge can be simply lifted off. Note also the raised pickguard and the raised fretboard. While not visible in this image, the neck is also at a slight angle to the body. There is no standard neck angle, but 4° - 5° is typical.

Some acoustic archtops are fitted with magnetic pickups. Since they are necessarily heavy, fixing them directly to the soundboard would reduce the volume and perhaps degrade the tonal quality. The solution for acoustic archtops has been to mount a pickup on the exposed end of the raised fretboard or cantilevered out from the pickguard; this is often called a floating pickup. Figure 6.29 shows a Washburn HB15TSK archtop guitar with such a pickup. Since this type of pickup is mounted far from the bridge – generally near the 21st or 22nd fret – there is reduced high frequency content in the resulting sound. This is roughly equivalent to turning up the tone knob. This 'soft' or 'mellow' sound is often preferred by jazz players.

Two other important distinctions in archtop guitars are the bracing pattern and whether the top is of solid or laminated wood. The two most common bracing patterns in archtops are a two-piece X brace and a two-piece, approximately parallel pattern. In either case, the braces are generally arranged so that they pass below the feet of the floating bridge, right under the height adjustment screws. The centers of these screws are typically about 3 in (76 mm) apart.

Fig. 6.28 An Au Naturel Model Archtop by Linda Manzer (Image courtesy of Linda Manzer, http://www.manzer.com, Photograph by Brian Pickell)

Fig. 6.29 A Washburn HB15TSK with a Floating Neck Pickup (Image courtesy of Washburn Guitars, http://www.washburn.com)

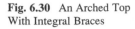
Fig. 6.30 An Arched Top
With Integral Braces

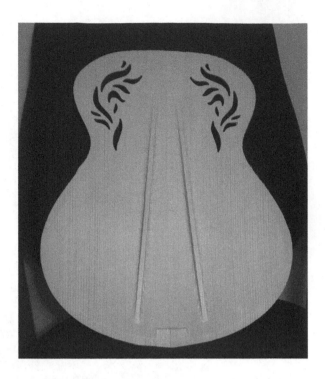

Figure 6.30 shows an experimental arched top (manufactured by Eddy Efendy, then a graduate student at Purdue University) using the parallel bracing pattern. Note that the term 'parallel' is only approximate since the braces are closer together at the top of the body than at the bottom.

Typically, braces are cut from separate pieces of wood and carefully fitted to the inside of the top plate. This is a time-consuming process, so this top was machined on a CNC router and has integral braces. There is a structural penalty since the grain of the braces is parallel to the centerline of the top rather than parallel to the centerlines of the braces themselves. This was partially corrected by adding a cap strip to the braces during assembly.

The final distinguishing characteristic of archtop guitars to be covered here is whether the top and back plates are solid wood or laminated. Traditionally, the tops and back plates are carved from book matched plates approximately 1.125 in (28.6 mm), a time-consuming process that also wastes the majority of the material. One obvious alternative is to form the top from veneer sheets and many instruments, particularly lower priced ones, use laminated tops. There doesn't appear to be any fundamental physical principle that would preclude making a very good sounding instrument using laminated wood.

Like other types of guitars, archtops have been the subject of experimentation. Figure 6.31 shows an arched top plate by Tom Stadler with an experimental bracing pattern consisting of a single center brace and several smaller lateral braces.

Fig. 6.31 Experimental Bracing Pattern for an Archtop Guitar (Image courtesy of Thomas Stadler, http://www.stadlerguitars.com)

6.5 Travel Guitars

A category of guitars that has been a source of innovation is travel guitars. With airline travel being such a routine activity – there are something like 600 million airline passengers per year in the US alone – there is a need for instruments that can be packed easily or stored in the overhead compartment on an airliner. Travel guitars are still a small portion of the overall market, but the number of choices seems to be growing.

Since travel guitars are generally smaller than conventional instruments, some are also being marketed for children. This has the added benefit of making instruments of generally better quality available for new, younger players. Too many instruments marketed specifically for children have been of such low quality that they might have frustrated otherwise enthusiastic new players. The increased availability of good, small instruments is welcome indeed.

Basic physics makes designing travel-sized acoustic instruments a challenge, since greatly reducing the size of the body affects the sound. There have been essentially two lines of development for travel-sized acoustic guitars. The first is to retain the scale length of a full-sized guitar and scale the body down separately. The second is to keep approximately traditional proportions while scaling the entire instrument down.

One of the most well-known travel guitars is the Martin Backpacker (Fig. 6.32). It was originally designed and manufactured by Bob McNally, but was licensed to Martin in 1992. Since then, more than 200,000 have been produced. McNally still produces stick dulcimers of similar design under the trademarked name Strum Sticks [3].

Fig. 6.32 Martin Backpacker
Travel Guitar (Image
courtesy of C.F. Martin
Archives, http://www.
martinguitar.com)

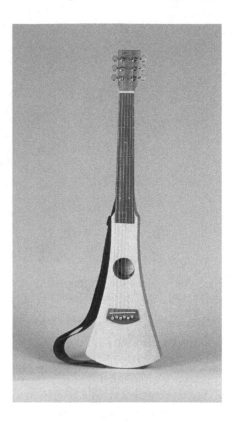

The Backpacker is about as compact as a playable acoustic instrument could be. The scale length is 24 in (610 mm), so it is at the lowest end of what could be called a full-sized scale length. The author has found the instrument to be quite playable, though the sound quality suffers from the small body volume and small radiating area of the top. That said, the instrument is well-suited to its purpose; it is easy to pack for travel and a good tool for individual practice or to accompany singing around a campfire.

Another, more recent design is the Washburn Rover (Fig. 6.33). It, too has a 24 inch (610 mm) scale length, but with a body more reminiscent of a conventional acoustic guitar. The author has found this instrument to have surprisingly good tone for such a small body. It is a bit larger than the Backpacker and is supplied with a rectangular case that fits easily into overhead bins on airliners.

A very popular example of travel guitars that approximately retain full-sized proportions is the Baby Taylor shown in Fig. 6.34. It is described by the manufacturer as a ¾ sized dreadnought and clearly has approximately the correct proportions. Note that a ¾ sized instrument is not actually 75% as large as a full scaled instrument. This nomenclature is commonly used for musical instruments, particularly for violins. The numerical size should be taken as qualitative rather than quantitative.

Fig. 6.33 Washburn Rover Travel Guitar (Image courtesy of Washburn Guitars, http://www.washburn.com)

Fig. 6.34 A Baby Taylor Travel Guitar (Image courtesy of Taylor Guitars, http://www.taylorguitars.com)

¼ size violins are widely available for young players and even 1/16 size instruments are manufactured.

The scale length on the Baby Taylor is 22.75 in (578 mm), near the low end for practical guitar scale lengths. While there are certainly exceptions, it is reasonable to consider 22 in (559 mm) as the lower practical limit for adult players. It is fair here to note that luthier Linda Manzer has made successful instruments with scale lengths as short as 43 cm (16.9 in). Figure 6.35 shows her Little Manzer guitar which has been played by jazz artist, Pat Metheny.

Table 6.8 shows dimensions of three representative travel guitars.

6.6 Artisan Guitars

We are fortunate to live at time in which many independent luthiers are working. Many produce fine guitars and some produce particularly innovative and artistic instruments. For lack of a better term, I refer to them here as artisan instruments. Builders whose instruments fall into this category often work alone or with a small

Fig. 6.35 A Very Small
Guitar Made by Linda
Manzer (Image courtesy of
Linda Manzer, http://www.
manzer.com)

Table 6.8 Dimensions of Travel Guitars

	Baby Taylor		Martin Backpacker		Washburn Rover	
	Inches	mm	Inches	mm	Inches	mm
Overall Length	33.75	857	33	838	33.5	851
Scale Length	22.75	578	24	610	24	609
Body Length	15.75	400			13.13	333
Body Width	12.5	318	7.25	184	8.25	210
Nut Width	1.69	42.9	1.69	42.9	1.69	42.9
String Spacing (Bridge)			2.187	55.5	2.125	54.0
Body Depth	3.375	85.7	1.94	49.2	2.0	50.8

number of assistants. Productions volumes vary, but 10-50 instruments per year is
not uncommon and in they are priced accordingly. This low volume, high skill
business model allows builders to work to a very high level and allows them to
carefully tailor instruments to the needs of their customers.

The group of luthiers working to a high level of craftsmanship and aesthetics is
too large to describe completely here. What follows is intended only as a hint of
what is available, and certainly not as a slight to those many who, only for the sake
of brevity, have been omitted.

Fig. 6.36 An Archtop Acoustic Guitar Designed and Built by Ken Parker (Image courtesy of Ken Parker, http://www.kenparkerarchtops.com)

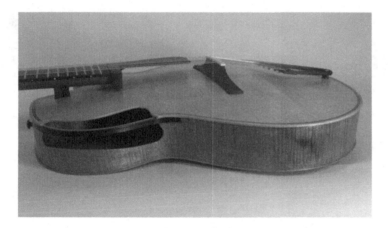

Fig. 6.37 Side View of An Archtop Acoustic Guitar Designed and Built by Ken Parker (Image courtesy of Ken Parker, wwwkenparkerarchtops.com)

A particularly interesting instrument is an archtop guitar made by Ken Parker, perhaps best known for his electric guitars (covered in the next chapter). This instrument, called the Semeur, uses a headstock reminiscent of his electric instruments (Fig. 6.36).

Apart from the obvious aesthetic appeal, perhaps the most interesting feature is the neck joint, visible in the side view in Fig. 6.37 and highlighted in Fig. 6.38. The designer's intent was to firmly attach the neck to the instrument, but in a way that made it easy to adjust and even to remove. As a result, the setup can readily be changed, the neck can be removed for stowage while traveling, and the instrument can be supplied with more than one neck [4].

Fig. 6.38 Front View of An Archtop Acoustic Guitar Designed and Built by Ken Parker (Image courtesy of Ken Parker, wwwkenparkerarchtops.com)

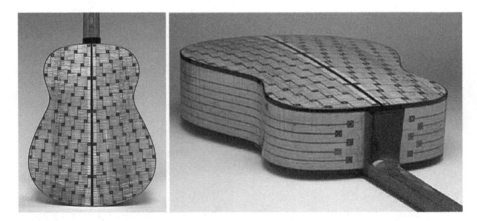

Fig. 6.39 An Elaborately Inlaid Acoustic Guitar by Ervin Somogyi (Image courtesy of Ervin Somogyi, http://www.esomogyi.com)

Particularly skilled builders sometime use their instruments as palettes for elaborate inlays or other decoration. Figure 6.39 shows a beautifully inlaid instrument by Ervin Somogyi, an accomplished luthier who has also written about guitar design [5-10].

References

1. Bruné R (2004) Eight Concerns of Highly Successful Guitar makers. American Lutherie 79:6–21
2. Brozman B (1993) The History and Artistry of National Resonator Instruments. Centerstream Publishing
3. http://www.strumstickcom, last accessed. on June 1, 2011

4. Personal conversation with Ken Parker, June 2, 2011
5. Somogyi E (1993) Priciples of Guitar Dynamics and Design? American Lutherie, #36
6. Somogyi E (2000) The State of the Contemporary Guitar: Part 1. Fingerstyle Guitar, #40
7. Somogyi E (2000) The State of the Contemporary Guitar: Part 2. Fingerstyle Guitar, #41
8. Somogyi E (2001) The State of the Contemporary Guitar: Part 3. Fingerstyle Guitar, #42
9. Somogyi E (2000) The State of the Contemporary Guitar: Part 4. Fingerstyle Guitar, #43
10. Somogyi E (2010) The Responsive Guitar. Hal Leonard Publishing

Chapter 7
Design Specifics for Electric Guitars

A wide variety of electric guitar designs are now in production, far too many to comprehensively discuss here. However, a large proportion is based on either Fender or Gibson designs. Design specifics for these two instruments will, thus, cover a range of instruments and serve as a good starting point for unrelated designs.

It's important to realize that designing a guitar, no matter how simple the final product may look, is the result of many design decisions. Especially in designs intended for large scale production, even small decisions can have serious implications. Designers, product managers and players debate choices of materials, hardware and structural geometry at far greater depth than I can show here. That said, a few basic dimensions drive the rest of the design, so it makes sense to start with them.

The most important dimension when designing a guitar is the scale length. In electric guitars, the most popular scale lengths are 25.5 in (648mm), associated with Fender, and 24.75 in (629mm), associated with Gibson. These are sometimes called long scale and short scale. Instruments intended for children or smaller players can have shorter scale lengths as do travel guitars designed to be easily transported and stored. Small scale electric guitars like these can have scale lengths as short as 22.75 in (575mm). One of the shortest scale lengths for a full sized instrument is 24 in (610 mm), used on the Fender Jaguar and Mustang [1]. Conversely, very few electric guitars use a scale longer than 25.5 in (648 mm).

Another primary design feature is the number of frets. Most solid body electric guitars have 21, 22 or 24 frets. Some makers (this one included) prefer 24 fret necks since that offers the player a two octave range – there are 12 semi-tones and thus 12 frets in an octave. While 21 or 22 frets do not offer a full two octave range, they do allow more space between the bridge and the end of the neck. Pickup location within this space strongly affects the amplified sound of the guitar and many designers consider placement to be more important than the number of available frets.

Figure 7.1 shows a solid body electric guitar made by the author with a 24 fret neck and a 25.5 in (648 mm) scale length. There is just enough room for two humbucking pickups and, since they are closely spaced, they do not offer quite as much tonal variation as would two more widely spaced ones.

R.M. French, *Technology of the Guitar*, DOI 10.1007/978-1-4614-1921-1_7,
© Springer Science+Business Media New York 2012

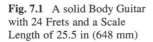

Fig. 7.1 A solid Body Guitar
with 24 Frets and a Scale
Length of 25.5 in (648 mm)

For comparison, a Fender Jazzmaster shown in Fig. 7.2 has the same scale length and its two pickups are roughly the same size as humbuckers. The pickups are clearly much farther apart, even though they are about the same distance from the bridge and neck respectively.

Another dimension that helps define a basic electric guitar design is the nut width. A typical nut width is 1 11/16 in (1.69in or 42.8 mm). Nut width is a surprisingly sensitive dimension. Changing width by 1 or 2 mm (around 1/16 in) is quite noticeable, particularly to an experienced player. A narrow neck for an electric guitar may be 1 5/8 in (1.625 in or 41.3 mm) while a wide one may be 1¾ in (1.75 in or 44.5 mm). It is rare for an electric guitar nut width to be outside this range.

With the scale length, number of frets and the nut width specified, another defining dimension is the body thickness. Strictly speaking the body only has to be thick enough to be structurally sound and to be comfortable. One of the thinnest electric guitar bodies is on the original Gibson SG at 1 5/16 in (1.3125 in or 33.3 mm) [2]. Fender guitar bodies are typically around 1.75 in (44.5mm) and this is a good average dimension for an electric guitar with a slab body [3]. At the other end of the scale, the Gibson Les Paul has an arched top that significantly increases the thickness of the body near the bridge. The maximum thickness of the body is 2 3/8 in (2.375 in or 60.33 mm)[4].

As discussed in an earlier chapter, there are basically three different ways of joining the neck to the body: Bolt on, set necks and neck through body. Of these,

Fig. 7.2 Fender Jazzmaster with Two Single-Coil Pickups (Wikimedia Commons, image is in the public domain)

only the bolt on has anything like standardized dimensions, the ones defined by the Fender Stratocaster and Telecaster and adopted by many other manufacturers.

The bolt on neck would be more correctly called a screw on neck since the fasteners are actually sheet metal screws going through the back of the body and into the back of the neck. The forces from the screws are usually distributed across the surface of the back of the body by a metal neck plate as shown at the top of Fig. 7.3. The neck plate size and hole spacing (at the bottom of Fig. 7.3) are widely used, even on instruments not made by Fender.

7.1 Representative Dimensions

Table 7.1 presents nominal dimensions for a solid body guitar with a bolt on neck. This is a very common configuration and may represent the majority of solid body electric guitars. One measurement not presented here is the angle between the neck and the body. The neck on many solid body guitars is parallel to the top of the body, but this is not universal. It is not uncommon for the neck to be angled back a few degrees [5], particularly on instruments with arched tops like the Les Paul. Finally, Table 7.2 presents typical string heights and neck relief.

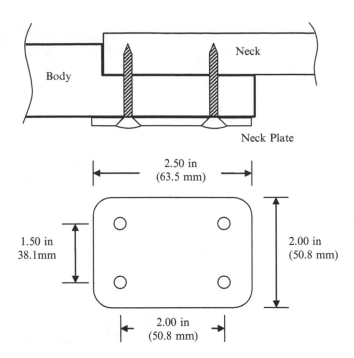

Fig. 7.3 Typical Dimensions for a Bolt on Neck

Table 7.1 Representative Dimensions for Electric Guitar with Bolt On Neck

Dimension	Value – inches	Value – mm
Headstock Thickness	0.50 – 0.54	12.7 – 13.7
Nut Width	1 5/8 – 1 3/4	41.3 – 44.5
Fretboard Thickness (at center)	1/4	6.35
Fretboard Radius	7.25 – 9.5	184 - 241
Neck Thickness at 1st Fret (including fretboard)	0.82	20.8
Neck Thickness at 12th Fret (including fretboard)	0.90	22.9
Neck Thickness at heel (including fretboard)	1.00	25.4
Neck Width at Heel	2 3/16	55.6
Depth of Neck Pocket	0.62 - 0.65	16.5
Length of Neck Pocket	3.00	76.2
Depth of Pickup Pocket	0.63	16
Body Thickness	1.75	44.5
Top Thickness at Control Pocket	0.200	5.1
Scale Length	24.75 – 25.5	629 – 648

Table 7.2 String Heights		String 6 – Bass E	String 1 – Treble E
for a Solid Body Electric Guitar	1st Fret	0.024" (0.61 mm)	0.010" (0.25 mm)
	12th Fret	0.078" (1.98 mm)	0.063" (1.60 mm)
	Relief	0.01" (0.3 mm)	at the 8th fret

Fig. 7.4 A Fender Telecaster with Front Electronics Pocket (Wikimedia Commons, image is in the public domain)

Because the range of both electronics and of body shapes is so large in solid body electric guitars, there is no standard electronics pocket. Indeed, there is not even agreement on whether the pocket should be on the front or back. Figure 7.4 shows a Fender Telecaster with both a large pickguard and an electronics pocket cover. These two covers hide the electronics pocket, edges of the neck pickup pocket and sometimes wiring channels.

Figure 7.5 shows an aftermarket Telecaster body before finishing and assembly. The pockets and wiring channels are clearly visible. This particular body has been made with no surface wiring channels. Rather, two separate holes have been drilled though the body. One runs from the neck pocket, through the front pickup pocket and into the electronics pocket. The other runs from the bridge pickup pocket to the electronics pocket. The two drilling paths are shown as dotted lines.

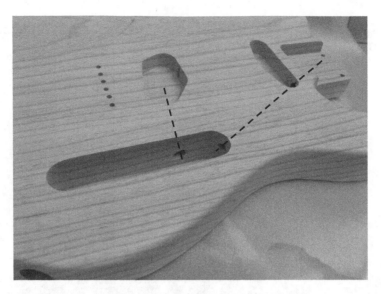

Fig. 7.5 An Aftermarket Telecaster Body with Internal Wiring Channels

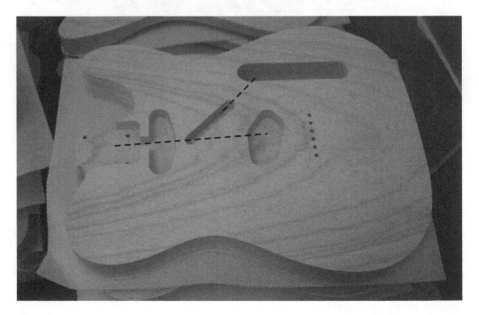

Fig. 7.6 An Aftermarket Telecaster Body Showing Wiring Channel (Image by the author, reproduced here courtesy of Wildwood Manufacturing, http://www.wildwoodmfg.com, drilling paths by the author)

Figure 7.6 shows an aftermarket Telecaster body by another manufacturer with a slightly different geometry. There is a gap between the neck pocket and the neck pickup pocket. Additionally, there is a diagonal channel between the two pickup pockets. There are no wiring holes drilled yet, but this design would make it easy to

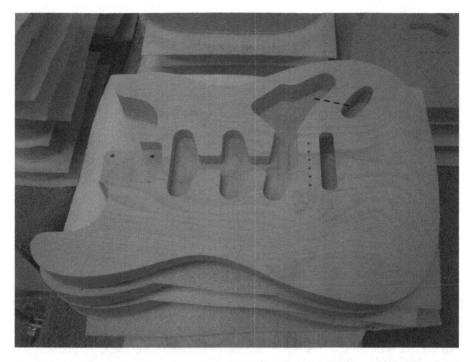

Fig. 7.7 An Aftermarket Stratocaster Body Showing Extensive Top Routing (Image by the author, reproduced here courtesy of Wildwood Manufacturing, http://www.wildwoodmfg.com, drilling path by the author)

drill a single straight hole from the neck pocket and the neck pickup pocket, through the diagonal channel and into the bridge pocket. A second short passage could be drilled from the bottom of the diagonal channel to the electronics pocket. These provisional paths are shown as dotted lines on the image.

A more extreme case is shown in an aftermarket Stratocaster body shown in fig. 7.7. Since the Stratocaster has a large pickguard, there is no problem with routing large channels in the top for both pickups and wiring. There is no need to drill additional holes in the body for wire passages other than for the wiring to the output jack (drill path shown by a dotted line). With the exception of the output jack, all the electronic components are mounted to the pickguard and become part of the body as soon as the pickguard is attached.

An even more flexible interpretation of the basic Stratocaster design in the aftermarket body is shown in Fig. 7.8. In this variation, the individual pickup pockets are replaced by a single long pocket and deeper side channels. This pocket can accommodate both single and humbucker sized pickups in any position and any order. The deep side channels are for the height adjustment screws.

In contrast to Fender designs, the Gibson Les Paul, intended from the beginning to be a more sophisticated instrument, requires more machining operations. The top is gracefully arched and pockets are machined in it for two humbucker pickups and

Fig. 7.8 An Aftermarket Stratocaster Body with a Universal Pickup Cavity (Image courtesy Stewart MacDonald, http://www.stewmac.com)

Fig. 7.9 A Gibson Les Paul (Wikimedia Commons, image is in the public domain)

well as holes for the bridge and tail stop mounting studs. The back generally has two separate pockets, one for the tone and volume controls and a smaller one on the top of the upper bout. This requires that the body be machined on both sides and that there be a wiring channel running diagonally, the full length of the body. Figure 7.9 shows a Les Paul with a figured maple top. The body is made of two different woods, maple for the top and probably mahogany for the back, so some of the internal channels can be routed before the top is glued to the back.

7.2   Electronics

Placing the electronics pocket on the front of the body requires some sort of cover, usually a plastic pick guard. The Fender Stratocaster design takes advantage of this requirement by mounting all the electronics directly to the pickguard so that it essentially becomes a module. The pickguard mounts a five position switch, three pickups, three potentiometers and associated wiring. Only the jack is not mounted to the pick guard. When the pick guard assembly is complete, it can be fitted into the body in a single operation. Figure 7.10 presents a wiring diagram for the Stratocaster that clearly shows how the various components are mounted directly to the pickguard.

Figure 7.11 shows the wiring diagram for a standard Telecaster. The basic wiring reflects its origin as a simpler precursor to the Stratocaster. Note that the two potentiometers and the switch all mount to the electronics pocket cover, forming a single module.

Finally, Fig. 7.12 shows the wiring diagram for a typical Les Paul. There are many versions of the Les Paul and some have different wiring configurations, but this one corresponds to the nominal design and is probably the most common.

The previous three wiring diagrams show three different pickup geometries and there is a fourth common one. The Stratocaster typically uses three single coil (or at least single coil sized) pickups and these often have a triangular base. A common variation uses the same coil geometry with an oval base plate. A third type of single coil package uses a flattened pentagon and is typically only used for Telecaster bridge pickups. The fourth common package size is for dual coil humbuckers.

Single coil pickups and humbucker pickups have almost standardized dimensions; there are small dimensional variations between different brands, but pickups of similar design from different vendors are generally interchangeable [6,7]. The dimensions shown in the following figures are drawn using published dimensions for several pickups by Seymour Duncan [8]. Figure 7.13 shows the nominal dimensions for a single coil pickup with an oval base and Fig. 7.14 shows the nominal dimensions for a single coil pickup with a triangular base. Note that there are many single coil sized pickups that are hum resistant. Some are actually dual coil humbuckers with smaller bobbins and others use more significantly different magnet and coil geometries.

Figure 7.15 shows the basic dimensions for a Telecaster bridge pickup. This, too, is a traditional single coil pickup, but hum resistant designs are also made to fit this packaging.

Finally, Fig. 7.16 shows nominal dimensions for a full sized humbucker pickup.

7.3   Archtop Guitars

One other type of electric guitar is an archtop with a center block. True acoustic archtop instruments have arched top and back plates without any connecting structure between them. However, they can have feedback problems on stage.

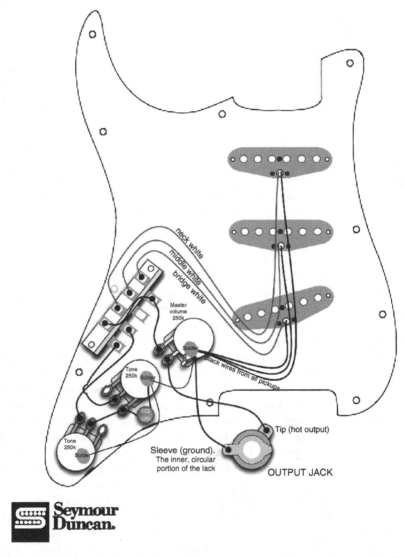

Fig. 7.10 Wiring Diagram for a Stratocaster (Image courtesy of Seymour Duncan, http://www. seymourduncan.com)

A solution has been to insert a solid block between the top and back, preserving the appearance and feel of an archtop guitar while largely eliminating feedback problems. One of the most popular of these designs is the Gibson ES-335 as shown in Fig. 7.17. The center block also provides a mounting structure for the bridge, tail stop and pickups. Of course, it also greatly limits top motion, so an

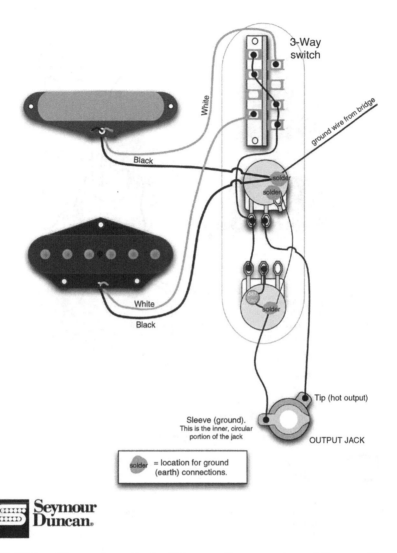

Fig. 7.11 Wiring Diagram for a Fender Telecaster (Image courtesy of Seymour Duncan, http://www.seymourduncan.com)

instrument of this design isn't practical as an acoustic instrument and must be played through an amplifier.

Electric archtop guitars with true hollow bodies generally don't mount the pickups directly to the top since the additional weight would greatly affect the resulting sound. Rather, they are usually mounted to the end of the neck or to the raised pickguard. Figure 7.18 shows a small humbucking pickup with tabs that allow it to be mounted at the end of the raised neck.

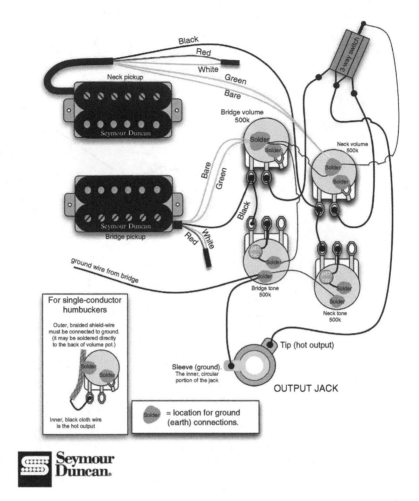

Fig. 7.12 Wiring Diagram for Typical Les Paul (Image courtesy of Seymour Duncan, http:// www.seymourduncan.com)

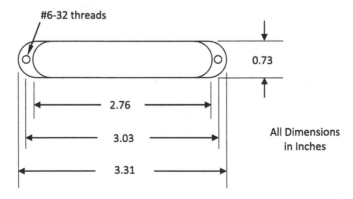

Fig. 7.13 Nominal Dimensions of a Single Coil Pickup with Oval Base

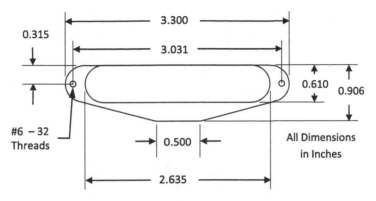

Fig. 7.14 Nominal Dimensions of a Single Coil Pickup with Triangular Base

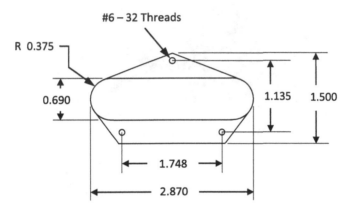

Fig. 7.15 Nominal Dimensions for a Telecaster Bridge Pickup

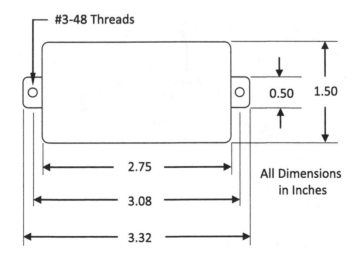

Fig. 7.16 Nominal Dimensions of a Humbucker Pickup

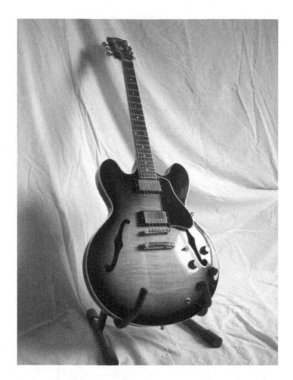

Fig. 7.17 A Gibson ES-335 (Wikimedia Commons, image is in the public domain)

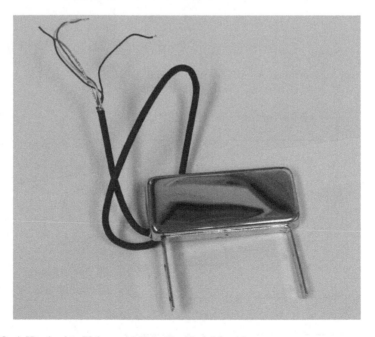

Fig. 7.18 A Humbucker Pickup with Tabs For Neck Mounting

References

1. Bacon T (2007) The Fender Electric Guitar Book. Backbeat Books
2. Bishop IC (1990) The Gibson Guitar. Bold Strummer
3. Hiscock M (2003) Make Your Own Electric Guitar. NBS Publications
4. Bacon T (2002) 50 Years of the Gibson Les Paul. Backbeat Books
5. Bacon T (2009) The Les Paul Guitar Book. Backbeat Books
6. Hunter D (2009) The Guitar Pickups Handbook. Backbeat Books
7. Milan M (2006) Pickups, Windings and Magnets:...And the Guitar became Electric. Centerstream Publications
8. http://www.seymourduncan.com. Last visited 25 july 2011

Chapter 8
Hardware

Almost any guitar includes some metal hardware such as tuning machines, a bridge, a truss rod and fret wire. Some instruments have significantly more than this. No discussion of guitar technology would be complete without a description of the hardware that is such an integral part of any instrument. For our purposes, let's consider hardware to be the things that are mounted onto the structure of the guitar. Except in rare instances, these are things the luthier purchases readymade.

8.1 Tuning Machines

Tuning machines have the necessary job of changing the tension of the strings to change their pitch – tuning. Essentially all guitars have tuning machines; the only type of guitar without them is the flamenco guitar – a light type of classical guitar with violin pegs instead of geared tuning machines. Tuning pegs serve the same purpose, but do so without the mechanical advantage offered by gears.

In mechanical terms, a tuning machine is a device that stretches a string to increase its tension and, thus, its pitch. It is a subtle difference, but it is important to note that a tuning machine applies a strain to the string, changing its length. It changes tension only by inducing a strain. While this may seem too picky a distinction to really matter, consider now what happens if an instrument with metal strings gets colder, perhaps by being placed in a cold car. The strings try to get shorter because of the decrease in temperature (the exact amount is determined by the coefficient of thermal expansion, α). Since the ends of the strings are fixed by the bridge at one end and the tuning machines at the other, the tension increases, raising the pitch. A tuning machine that directly changed tension would keep the string in tune, even with temperature changes.

R.M. French, *Technology of the Guitar*, DOI 10.1007/978-1-4614-1921-1_8,
© Springer Science+Business Media New York 2012

Fig. 8.1 A Wooden Worm
Gear Made for
Demonstrations (Wikimedia
Commons, image is in the
public domain)

Almost all guitar tuning machines are built around worm gears. The worm gear is a very common mechanical element that has been in use for a very long time. They are particularly useful for guitar tuning machines because they can have a large reduction ratio – guitar tuners have as much as 18:1 – and because they aren't reversible – the tension of the string cannot force the gear to turn backwards. Figure 8.1 shows a wooden model of a worm gear from the Museum of Arts and Crafts in Paris.

Figure 8.2 shows the large geared tuning machines on the back of a bass guitar headstock. Note that the gears are towards the body and the shafts with the tuning knobs are toward the end of the headstock. This is the case with all worm gear tuning machines.

Usually, tuners are made as individual units, though sometimes several are mounted to a single plate. It is traditional for classical tuners to be mounted three to a plate as shown in Fig. 8.3.

There are a few distinctions that can be used to sort out the different kinds of guitar tuning machines. One is whether they are intended for steel or nylon strings. Classical guitars originally used tuning pegs held in place by friction. These are the same type of pegs used in violins and members of the violin family and are shown in Fig. 8.4. Classical guitars now almost all use geared tuning machines as shown in Fig. 8.3, while flamenco guitars still sometimes use pegs.

A distinguishing feature in all geared tuners is the gear ratio. Most tuners have gear ratios in the range of 12:1 to 18:1. For example, a gear ratio of 14:1 means that the knob turns 14 times in order for the string post to turn once. Higher gear ratios allow more precision in tuning the strings.

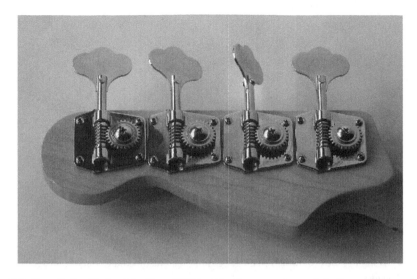

Fig. 8.2 Bass Guitar Tuning Machines (Wikimedia Commons, image is in the public domain)

Fig. 8.3 A Classical Guitar Headstock (Wikimedia Commons, image is in the public domain)

Another, less obvious, feature of tuning machines is how they are mounted. Classical tuners intended for nylon strings and some tuners designed for steel strings are mounted three to a plate. Figure 8.5 shows the headstock of a 1927 Martin 000-18 with three-on-a-plate tuners and a slotted headstock. These tuners are very similar to the classical tuners in Fig. 8.3 except that the posts are solid metal and smaller diameter.

Fig. 8.4 Violin Peg Box Showing Friction Pegs in Place (Wikimedia Commons, image is in the public domain)

Fig. 8.5 Headstock of a 1927 Martin 000-18 Guitar With Plate-Mounted Tuners (Image courtesy of C.F. Martin Archives, http://www.martinguitars.com)

Fig. 8.6 Tuning Machines for Solid Headstocks

Tuning machines designed for solid rather than slotted headstocks are generally mounted individually through holes drilled vertically through the headstock. They also have rings, called ferrules, that help keep the posts located correctly. There are essentially two types of ferrule mounting as shown in Fig. 8.6. The simpler models built around a flat plate have ferrules that press in from the top. Tuners made using a cast body typically have a barrel with internal threads. The ferrules have external threads and are installed with a washer to distribute the pressure against the surface of the wood.

The final distinguishing feature about tuning machines is how they are fixed to the headstock. Even though a threaded ferrule can be tightened enough to firmly hold the tuning machine in place, some sort of connector is needed to keep it from rotating due to the torque of the string wrapped around the post. Tuners are secured either by small screws or by two pins cast into the body of the tuner as shown in Fig. 8.7.

Before leaving the subject of tuning machines, it is worth pointing out that there are a small number of unconventional designs on the market at this writing. Perhaps the most interesting are from Ned Steinberger, a designer well known for instruments with minimalist bodies and no headstocks. Figure 8.8 shows bassist Victor Wooten playing a Steinberger bass. Steinberger guitars and basses are almost all headless and the tuners are integrated into the tail stop.

The tuners are built around large, fine-thread screws with knurled ends that can be turned by hand to pull on a block holding the ball end of the string. Many, but not all, of these instruments are designed to use strings with ball ends on both ends of the strings. Figure 8.9 shows a patent drawing for this type of tuner.

Fig. 8.7 Tuning Machines Showing Two Different Methods of Preventing Rotation

Fig. 8.8 Victor Wooten Playing A Steinberger Headless Bass (Wikipedia Commons, image is in the public domain)

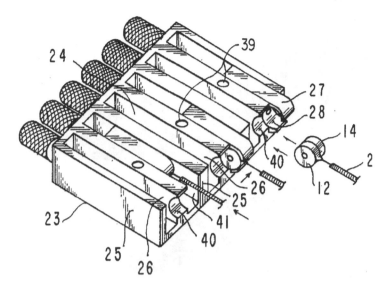

Fig. 8.9 Drawing From Ned Steinberger Patent for Body Mounted Tuning Machines (Patent 4,608,904)

Another interesting type of tuner, also from Steinberger, is a gearless one designed as a replacement for conventional worm gear tuning machines. Rather than winding the string around a peg that rotates as the knob is turned, these tuners directly stretch the strings by pushing the ends down into a threaded barrel. The threads have a very fine pitch to allow for precise changes in tension. Figure 8.10 shows a set of Steinberger gearless tuners fitted to the headstock of a Gibson Firebird.

8.2 Bridges

The obvious purpose of a bridge is to form the end points of the six strings. For almost all guitars, the bridge also provides the correct intonation (length correction) for the strings. On most acoustic guitars, the saddle is made in one piece and the position of the bridge is fixed. On most electric guitars, there are individual, adjustable saddles for each string. The two types of bridges are different enough that it makes sense to describe them separately.

8.2.1 Acoustic Bridges

Bridges on acoustic guitars need to be light to allow the top to vibrate. As a result, they are generally made of wood. They are either glued to the top or are held in place by the force of the strings. Those held in place only by string force are called floating bridges.

Gibson Firebird Reissue

Fig. 8.10 Steinberger Gearless Tuners (Image courtesy Stewart MacDonald, http://www. stewmac.com)

Figure 8.11 shows the most common type of acoustic guitar bridge. It is glued to the top and the strings are held in place using bridge pins. The saddle is slightly angled so that the strings are intonated reasonably well. Note also the slight offset in the saddle for the B string.

Fixed acoustic bridges are generally glued on after the body is finished. Typically, the finish under the saddle is scraped away so that glue can bond with the wood. The gluing area under the bridge in this picture is on the order of 4200 mm^2 (6.5 in^2) and the string tension is on the order of 445 N (100 lb). Thus, the glue needs to withstand a stress of about 106 kPa (15.4 psi). This is well within the strength limits of even a weak glue. The web site for Titebond® wood glue, a good quality wood glue that has been widely used in guitar making, [1] lists the bond strength as 3,600 psi (24.8 MPa).

Taylor Guitars uses an elegant way to mask off the soundboard where the bridge is to be glued on. Before the finish is applied, a clear plastic mask is fixed to the top precisely where the bridge is to be placed. The mask is a little smaller that the bridge and the bridge has a very light relief around the edge to accommodate the thin layer of finish and still sit firmly on the bare wood. This approach requires a very precise assembly process – a Taylor hallmark.

After the finish has been applied very precisely by an industrial robot and cured, the plastic mask is carefully lifted to expose the bare wood underneath. The bridge is then glued in place. Figure 8.12 shows a Taylor body made of figured koa. The clear mask is still in place and the wood underneath is lighter because the finish hasn't penetrated there.

Fig. 8.11 Acoustic Bridge Showing Angled Saddle and Bridge Pins

Fig. 8.12 A Taylor Guitar Body with the Bridge Mask in Place (Image by the author, reproduced here courtesy of Taylor Guitars, http://www.taylorguitars.com)

A very nice example of a guitar with a floating bridge is the Benedetto 16-B shown in Fig. 8.13. This instrument uses a very traditional wood bridge with threaded rings for height adjustment. Because it is held in place only by string tension, it can be moved for precise intonation. Floating bridges are usually found

Fig. 8.13 Benedetto 16-B Archtop with Floating Bridge (Image courtesy Bob Benedetto, http://www.benedettoguitars.com)

Fig. 8.14 An Acoustic Bridge with Adjustable Saddles (Image courtesy of Brian Yarosh, http://www.castorinstruments.com)

on archtop guitars. Since they must fit well in order to mechanically couple to the top, it is important that the feet of the bridge very accurately fit the curved top.

Figure 8.14 shows an interesting bridge by Brian Yarosh [2]. It has six individually adjustable saddles and a stop block so that bridge pins are not needed.

Fig. 8.15 A Simple Fixed Bass Bridge (Wikimedia Commons, image is in the public domain)

8.2.2 Fixed Electric Bridges

Bridges for electric guitars serve the same purpose as those for acoustic guitars – forming an accurate end support for the strings. However, electric bridges are generally much heavier and offer several adjustments. The weight is partly a byproduct of the mechanical complexity, but also serves to increase sustain. Since the sound is produced by the pickup sensing string motion, there is no need to let the kinetic energy of the strings vibrate the body as on an acoustic guitar. If the bridge is heavy compared to the strings, mechanical impedance is high and little kinetic energy is lost to the bridge.

There are many different electric bridge designs so it helps to have some system of organizing them. One distinguishing characteristic is whether they are fixed to the guitar or can be moved with a lever to change string tension quickly, changing pitch. The musical definition of this effect is vibrato, but the devices that produce it on guitars are often called tremolos. Strictly speaking, tremolo is a change in volume, not pitch, but the terms are sometimes used interchangeably [3].

Another obvious distinction is how the bridges are mounted to the body. Almost all bridges are either fixed to the body with screws or fixed to studs pressed into holes drilled in the body. Stud mounted bridges often have a separate tailpiece in addition to the adjustable bridge.

One of the most basic kinds of electric bridge is shown in Fig. 8.15. This is a bass bridge, but many guitar bridges are similar. Telecaster bridges, in particular, are often made this way.

The other common style of fixed bridge is a two piece design in which the rear piece is a tail stop for the strings and the forward piece forms a mount for six adjustable saddles. The original bridge using this design was introduced in 1954 on

Fig. 8.16 Tune-o-matic Bridge with Tail Stop (Wikimedia Commons, image is in the public domain)

the Gibson Les Paul Custom and was called the Tune-o-matic. Figure 8.16 shows a Tune-o-matic bridge on a Les Paul.

The two components are mounted on studs set into the body of the guitar. The studs for the tail stop are heavy because they have to resist the tension of the strings. Those for the bridge itself can be lighter since they only have to withstand the vertical component of the string tension due to the break angle over the bridge. Note that the bridge studs include thumbwheels that allow the height of the bridge to be adjusted easily. Figure 8.17 shows a Tune-o-matic bridge along with the internal components.

A design related to the Tune-o-matic is the stud mounted wraparound bridge. This design merges the tailpiece and saddles into the same component. There are several different models currently available, two of which are shown in Fig. 8.18. The bridge on the right is the simplest type of wraparound design. The saddles are cast into the top of the bridge and there is a set screw in each of the two stud mounting slots. The set screws allow the entire bridge to be moved with respect to the studs,

The bridge on the left has an adjustable insert with two saddles cast into it. The position of the G and B saddles can be changed, though only together. Note the heavy studs also shown in Fig. 8.18. Each has a heavy brass barrel with vertical grooves on the outside and threads on the inside. The barrels are pressed into holes in the guitar body so that they can form a solid mounting point for the studs. The chrome plated studs can be used to adjust the height of the bridge while fixing it to the guitar. While there is no defined standard for stud spacing, they are generally 3.25 in (82.6 mm) on center.

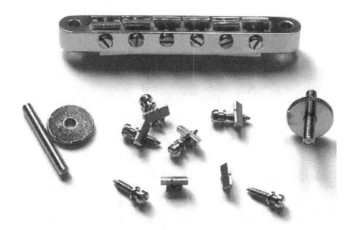

Fig. 8.17   Tune-o-matic Bridge with Parts (Wikimedia Commons, image is in the public domain)

Fig. 8.18   Two Types of One-Piece Wraparound Bridge

8.2.3   *Electric Tremolo Bridges*

Not all electric guitar bridges are fixed. Rather, many of them are designed to allow the player to add vibrato by using a lever to quickly change the tension on the strings. In spite of their effect being vibrato, they are usually called tremolo bridges.

Tremolo Bridge with No Force on Whammy Bar

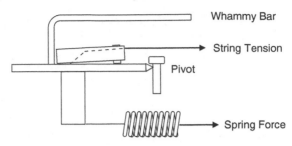

Tremolo Bridge with Downward Force on Whammy Bar

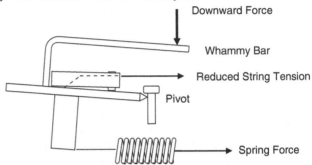

Fig. 8.19   Operation of a Tremolo Bridge

There are a number of different designs available, but they mostly fall into one of three types: Fender, Bigsby and Floyd Rose.

The Fender tremolo bridge was introduced on the Stratocaster and has been used on it ever since. There have been a small number of Strats with fixed bridges – generally known as hard tail Strats – but a large majority of Strats have tremolo bridges.

The operation of the Stratocaster bridge is very simple. The bridge itself pivots against a set of screws (as few as two or as many as six) set into the body of the guitar. The tension of the strings at the top of the bridge assembly is countered by springs connecting the bottom of the bridge to the body. The two torques counteract one another so the strings stay at the correct pitch until an additional torque is provided by the player pushing or pulling on a lever connected to the bridge. This lever is usually called a whammy bar. Figure 8.19 shows the basic operation of a tremolo bridge.

Figure 8.20 shows a close-up view of the tremolo on the author's Squire Standard Stratocaster. The two black screws on which the bridge pivots are clearly visible at the front of the bridge. Note also the whammy bar in place. Many Strats are played without the whammy bar installed, so it is not at all unusual to see a hole at the bottom of a Strat bridge. Some players even choose to 'block' their bridges by inserting a block (usually wood) in between the bridge and the body (usually on the back of the instrument). This prevents the bridge from moving, even under pressure of the player's hand.

Fig. 8.20 Stratocaster Tremolo Bridge with Whammy Bar in Place

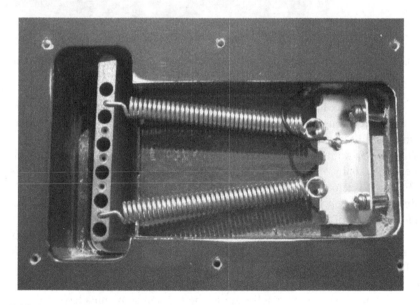

Fig. 8.21 Back of a Stratocaster Tremolo Bridge Showing Springs

Figure 8.21 shows the back of the author's Stratocaster with the cover plate removed. There are two springs that counter the string tension. Note that there are five tabs on the spring mount and five holes in the bottom of the bridge. It is not unusual to see Strats with five bridge springs.

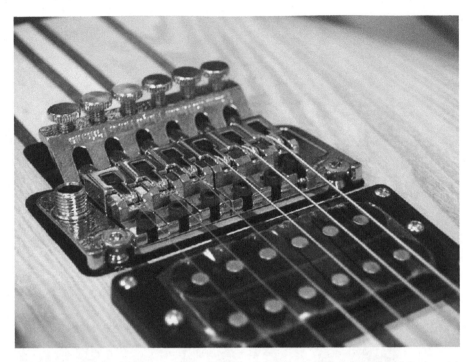

Fig. 8.22 A Floyd Rose Tremolo Bridge (Wikimedia Commons, image is in the public domain)

Note also the six large holes in the bridge. The strings are loaded into the bridge from the back through these holes and the rear cover plate has openings that allow the strings to be changed without removing the cover.

The Fender tremolo bridge has worked well for decades, but there have been refined versions by other manufacturers. Probably the most well known of these is the Floyd Rose locking tremolo, an example of which is shown in Fig. 8.22. Designed by the eponymous Floyd in the late 1970s, this bridge was an attempt to address some of the limitations of the traditional Fender tremolo.

As the player pushes on the whammy bar and rotates the bridge around its pivot points, the string tensions change. Since the strings are elastic and stretch under load, the change in tension slightly changes the length of the strings. This means that the strings need to slide slightly over the nut. If they do not slide freely, they can go out of tune. Some players apply a lubricant to the nut and some nuts have small roller bearings that reduce the friction of the strings. These are usually called roller nuts.

Rose solved this problem by adding clamps to the nut so that the strings are fixed at the nut rather than the tuners. The strings are tuned normally with the clamps loosened and then the clamps are tightened to lock the strings to the nut. The bridge also incorporates locking features, so the player can create significant pitch changes and still have the strings go back to proper pitch. Because of the string locking components at both ends, the bridge is called a double locking bridge.

Fig. 8.23 A Gretsch Archtop Guitar with a Bigsby B6 Vibrato Tailpiece (Image courtesy of Bigsby Guitars, http://www.bigsby.com)

There are a number of different versions of the Floyd Rose bridge in use, including ones licensed by manufacturers. The one shown in Fig. 8.22 includes fine tuners, fine thread screws that slightly change the string tension by moving the string anchors. They allow the player to make small tuning adjustments without having to unlock the clamps at the nut.

The other major type of pitch changing device is the Bigsby tailpiece as shown in Fig. 8.23. It should be noted that Bigsby correctly calls their products vibrato tailpieces. There are different Bigsby designs to fit different instruments. The B6 model shown here is designed to replace the tailpiece on an archtop instrument. The strings are wrapped around a shaft and held in place by pins along the bottom of the shaft that go through the centers of the ball ends. Pushing up or down on the handle (whammy bar) rotates the shaft, changing tension in the strings.

The Bigsby is designed to work with a Tune-o-matic type bridge and replaces the fixed tailpiece. The strings are attached through a circular tube that rotates as the whammy bar is pushed, thus, changing the tension.

8.3 Truss Rods

Almost all guitars have some sort of reinforcing rod in the neck. The exceptions are classical and flamenco guitars. Truss rods have two purposes: to stiffen the neck and to change its curvature. The tension of the strings tends to curve a neck upwards (in engineering convention, this is positive curvature). Players usually prefer a

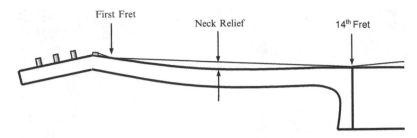

Fig. 8.24 Measuring Relief in an Acoustic Guitar

Fig. 8.25 An Electric Guitar Neck with a Single Acting Truss Rod (Wikimedia Commons, image is in the public domain)

small amount of positive curvature (called relief in the guitar world). The traditional way of measuring relief is to press a selected string to two reasonably separated frets. The neck crosses the body at the 14th fret on most steel stringed acoustic guitars, so the 1st and 14th frets are a good choice. The distance between the bottom of the string and the top of the fret nearest the middle of the ones being used for the measurement is the relief (Fig. 8.24). Neck relief on the order of 0.25 mm (0.010 in) is typical. When a fixed truss rod is used, the builder either has to either assume that the neck will bend slightly under the tension of the strings or has to sand the desired relief into the fretboard.

A much more flexible solution is to use an adjustable truss rod. It allows the curvature of the neck to be adjusted to suit the player. Adjustable truss rods can be divided into two categories, single acting and double acting. As the name suggests, a single acting truss rod can induce curvature in only one direction. Usually, a single-acting truss rod is essentially a long, thin steel rod with a fitting on one end to secure it to the neck and threads on the other end. The threads allow a nut to be used to adjust tension and, thus, the curvature of the neck. Figure 8.25 shows a cross-section of an electric guitar neck with a single acting truss rod.

Note that the truss rod is not straight. Rather, it is curved upward on the ends (positive curvature). When the tension is increased, the center of the neck is raised and the relief is decreased. This type of truss rod is capable only of reducing neck relief. The assumption when using this installation is that the string tension will create more neck relief than desired and that the truss rod is used only to reduce the relief.

Another type is the double-acting truss rod. This type is capable of inducing both positive and negative curvature so it can increase or decrease neck relief. Figure 8.26 shows a representative example, patterned after a Hot Rod by Stewart MacDonald. It consists of two threaded rods and two threaded blocks. The lower rod has an Allen nut at one end for adjustment.

Fig. 8.26 Representative Double Acting Truss Rod

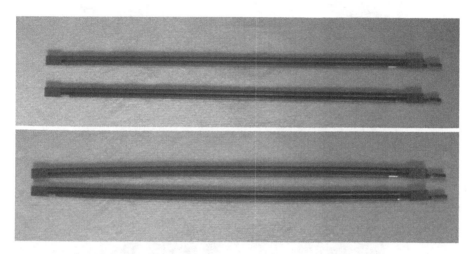

Fig. 8.27 Double Acting Truss Rods without (top) and with Curvature (bottom)

This type of double acting truss rod works because of the differential between the top and bottom threaded blocks. When the nut is turned one way, the lower rod tries to push the blocks farther apart. When the nut is turned the other way, it tries to pull them together. In either case, the top rod does not rotate and acts to keep the blocks from moving. Thus, in either case, the net result is a curvature in the rod. Figure 8.27 shows two double acting truss rods straight and the same two rods with opposite curvatures.

There are several common ways of installing a truss rod in a neck. It is important to remember that there has to be access to one end of the rod so that it can be adjusted. Figure 8.28 shows possible installations of a double acting truss rod in electric and acoustic necks.

Often, truss rods in acoustic guitars are installed so that the adjusting nut is at the heel and is accessed through the soundhole. Taylor acoustic guitars currently use a bolt-on neck design with the adjustment nut at the headstock. Figure 8.29 shows a Taylor neck during production with the adjustment nut exposed. In the finished instrument, the adjustment nut will be hidden with a decorative cover.

In almost all cases, truss rods are set into slots cut or routed into the neck. Often, the slot is cut into the neck before the fretboard is glued on. Figure 8.30 shows a neck with the truss rod slot, but fretboard not installed yet.

Another popular way to install the truss rod is from the back. In necks where the fretboard and neck are all one piece, this is a convenient way to place the truss rod. Fender necks that have the truss rod installed from the back typically have a

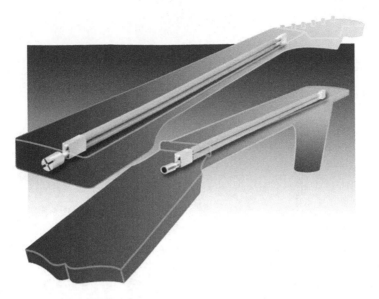

Figure 8.28 Sample Installations of a Double Acting Truss Rod (Image courtesy Stewart-MacDonald, http://www.stewmac.com)

Fig. 8.29 A Taylor Acoustic Neck Showing the Truss Rod Adjustment Nut (Image by the author, reproduced here courtesy of Taylor Guitars, http://www.taylorguitars.com)

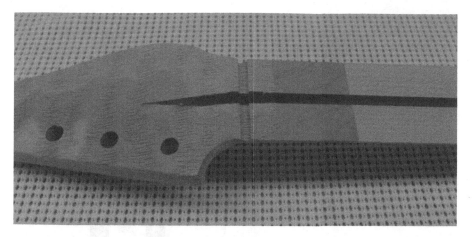

Fig. 8.30 A Guitar Neck with Truss Rod Slot

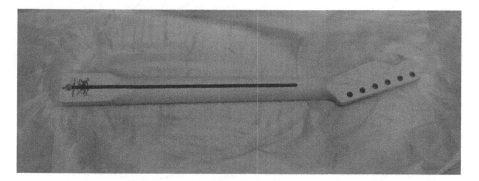

Fig. 8.31 An Aftermarket Telecaster Neck with a Skunk Stripe

contrasting piece of wood filling the pocket through which the truss rod is installed. This is colloquially called a skunk stripe. Figure 8.31 shows an aftermarket maple neck from Mighty Mite (licensed from Fender) with a dark skunk stripe.

Fixed truss rods are generally either made of steel or graphite and may be solid rectangular bars or rectangular tubes. They are typically set into slots in the neck under the fretboard; a typical neck cross-section is shown in Fig. 8.32.

Occasionally, guitars are made with more than one truss rod in the neck. One configuration uses a combination of fixed and adjustable truss rods as shown in Fig. 8.33. This configuration results in a stiffer neck than one with just an adjustable rod. However, the additional stiffness provided by the fixed rods limits the deflection that can be induced with the adjustable rod.

The other multi-rod configuration occasionally used in guitar necks is two adjustable truss rods as shown in Fig. 8.34. This allows the possibility of both changing the curvature of the neck and inducing (or correcting) a slight twist.

Fig. 8.32 Typical Neck
Cross Section Showing a
Fixed Truss Rod

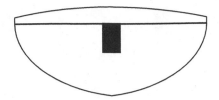

Fig. 8.33 Neck Cross-
Section Showing Two Fixed
Truss Rods and One
Adjustable one

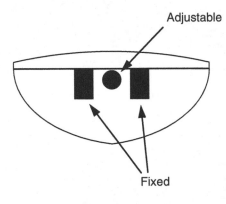

Fig. 8.34 Neck Cross-
Section Showing Two
Adjustable Truss Rods

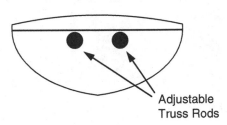

If tension is identical in the two rods, the curvature of the neck is changed. If the tension is different, a twist is induced along with some curvature. In order to induce only a twist, the truss rods would need to be double acting so that they could produce equal and opposite internal forces.

References

1. http://www.titebond.com. Web site last visited June 3, 2011
2. Yarosh B (2005) Adjustable Saddle for Acoustic Guitars. American Lutherie #82
3. Kennedy M and Bourne J, ed. (2006) The Oxford Dictionary of Music, 2nd ed. Oxford University Press.

Chapter 9
Iconic Guitars

The number of existing guitar designs is quite large, but a small number of them might be considered iconic. It's a risky thing to attempt to define what an iconic instrument might be. It is perhaps riskier still to select a small group of instruments to be included in such a list. Nevertheless, what follows is an attempt to do just that. The list of instruments and their descriptions appear here to show some of the most important developments in the history of the guitar. These are instruments that were popular enough to have affected the design of those that followed them. This is not intended to be an authoritative collection. Indeed, if it sparks a debate about instruments that should or should not have been included, then this author is happy.

The history of these instruments is well-documented elsewhere; the descriptions here cover their importance to the design and development of the guitar.

9.1 Torres Classical Guitar

The Torres classical guitar from 1856 essentially fixed the design for the modern classical guitar. Certainly, there were other accomplished builders who were contemporaries of Torres, but his instruments did much to establish design features that are now common to most classic guitars. Not all the design elements of Torres' guitars were new at the time, but these instruments were the first to incorporate them all at once [1].

The collection of modern features that would be familiar to Torres is large; perhaps we can start with overall proportions. Figure 9.1 shows the soundboard from a modern set of plans for an 1864 Torres, drawn by Neil Ostberg [2]. Though the original instrument is now more that 150 years old, it looks very familiar. The body width is approximately 354 mm (13 15/16 in) and the body depth is

R.M. French, *Technology of the Guitar*, DOI 10.1007/978-1-4614-1921-1_9,
© Springer Science+Business Media New York 2012

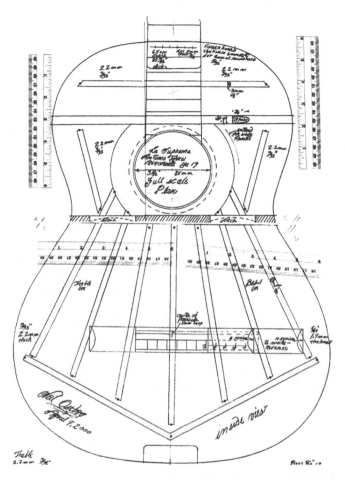

Fig. 9.1 Plan Sheet Showing Top of Torres Guitar (Image courtesy of Neil Ostberg, mysite. ncnetwork.net/resr7g3w)

approximately 99 mm (3 29/32 in). The scale length is approximately 654 mm (25 ¼ in). Indeed, just these three numbers are enough to establish the overall size of the instrument.

The plan sheet clearly shows the fan bracing, the rectangular bridge with tie block and other internal elements. Not only are some modern instruments faithful copies of Torres designs, but many incorporate features from these instruments. Figure 9.2 shows a modern classical guitar made by Stephen Marchione. While it is not a direct copy of a Torres guitar, the design is a very traditional one and there are broad similarities.

Another interesting example of the influence of Torres is an instrument made in 1949 by Robert Bouchet. Figure 9.3 shows the body of this instrument with the back removed. Compare the bracing pattern of this instrument with that shown in the plans in Fig. 9.1.

Fig. 9.2 A Classical Guitar
Made by Stephen Marchione
(Wikipedia Commons, image
is in the public domain, see
also Marchione.com)

9.2   Martin OM-18/28

The story of the flat top, steel string acoustic guitar is, in large part, the story of Martin Guitars. C.F. Martin started making guitars in the United States in the 1830s. He left Saxony (a state in what is now Germany) with his family in 1833 and, by 1834, had established an instrument retailing and manufacturing business in New York City. The Martin Company has made a large variety of guitars since its founding, but a few stand out as being truly iconic. One of these is the OM-28 and the less expensive version called the OM-18.

The OM-28 has been described by Dick Boak, Director of Artist Relations for Martin Guitars, author and Martin historian, as the first truly modern Martin Guitar [3]. Earlier Martin designs were gut-stringed instruments that had been modified to withstand the increased tension of steel strings. Some had slotted headstocks and hardly any of them had pick guards.

However, the OM-28 was designed from the start to be a steel stringed instrument and included a unique combination of features. The body shape was modified from the Martin 000, shortened so that the neck crossed the body at the 14th fret rather than the 12th fret. The neck was longer and slimmer than in previous designs and a pick guard was added. The headstock was solid and had banjo-style tuners. A characteristic internal feature is that it used X-bracing.

Fig. 9.3 Interior of a
Classical Guitar Made in
1949 by Robert Bouchet
(Image courtesy of
Richard Bruné, http://
www.rebrune.com)

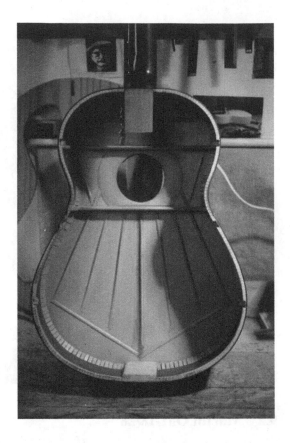

Some of the features in the OM-28 were used in other instruments. For example, the Gibson L-1, a flat top acoustic guitar first introduced around 1926, used an x-bracing pattern after the first year or so of production. Similarly, other instruments, such as the Gibson L-5 archtop, had 14 fret necks. The OM-18/28 design became iconic, at least in part, because it brought the right combination of features together in one instrument.

The instrument was originally marketed in 1929 as the 000-28 Orchestra Model since the body shape was a modification of the Martin 000 body. This was quickly shortened to OM-28. The OM-28 had rosewood sides and back. When the onset of the Great Depression suggested the need for a more affordable instrument, a less expensive mahogany version was introduced in 1930 as the OM-18. Fig. 9.4 shows the components of a partially completed OM-28 replica made by Jim Merrill. The familiar X-bracing is clearly visible on the top as are the parallel cross braces on the back.

Figure 9.5 shows luthier Jim Merrill shaping the braces on an OM-28 replica. Note that the braces reach their maximum height near the center of the soundboard just below the soundhole. Recall that stiffness of a beam is a function of thickness cubed (raised to the third power), so the bracing greatly stiffens the soundboard at this point[4].

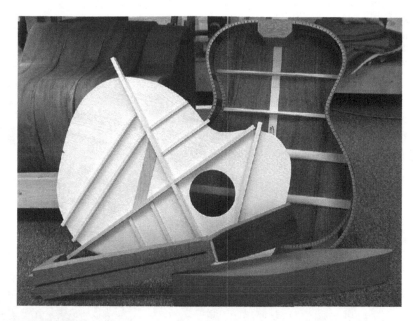

Fig. 9.4 A Partially Completed Martin OM-28 Replica Made by Jim Merrill (Image courtesy Jim Merrill, Merrillguitar.com)

Also, note that the thickness of the braces is reduced to essentially zero at the edges of the soundboard. While the soundboard has to be stiff enough to resist the pull of the strings, it also needs to be flexible enough to move in response to the vibration of the strings. Thus, it is important that the stiffness of the soundboard be as low as possible at the edges.

As an indication of how influential the OM-28 design has been, consider the instrument in Fig. 9.6. This is a Taylor model 114, a popular model from one of the most successful guitar companies. It, too, has a solid headstock, a 14 fret neck, an X-braced top, a parallel braced back and a pick guard. The most notable external differences are aesthetic - unique shapes for the headstock, bridge and pick guard. Indeed, these aesthetic differences are important in distinguishing different brands and are carefully protected proprietary designs.

For all its similarities, it is not at all correct to think of the Taylor 114 as just a slightly modified copy of the Martin OM-28. It certainly represents a major effort by Taylor, and many of its unique features are not visible from the outside. One of these is a bolted neck joint that is much more sophisticated than glued dovetail neck joint used on the original OM-28.

The Taylor guitar is made using different materials as well. While the top is still spruce, the sides and back are laminated from sapele, a wood native to tropical Africa with a material properties similar to mahogany. The glue is a modern aliphatic resin (yellow wood glue, e.g. Titebond®) and the finish is ultraviolet-cured polyester. It should be noted that modern versions of the OM-28 produced by Martin also include significant refinements from the original design.

Fig. 9.5 Luthier Jim Merrill
Carving the Braces on an
OM-28 Replica (Image
courtesy of Jim Merrill, http://
www.merrillguitars.com)

Perhaps the most important feature of the 114 is one not visible at all on or in the guitar itself. Taylor has been at the forefront of efforts to bring modern, computer-controlled production methods to guitar making [5]. The uniformity and precision of the resulting instruments is well beyond anything possible in 1930. Even the finish is sprayed and buffed using industrial robots.

For all the innovations on the Taylor 114 and its manufacturing processes, it is immediately familiar to anyone used to looking at the Martin OM-28. It is probably fair to think of it as a modern descendent of the older Martin.

9.3   Martin D-28

A quick look at steel string acoustic guitars shows that bodies are generally of two different forms. One is the curvy, 'hour glass' shape typified by the Martin OM-28 and its relatives. The other is a more angular body shape called a dreadnought. Martin first started making the precursors of the now-familiar dreadnought guitar starting in 1916. It made the guitars for the Oliver Ditson Company and used the Ditson body shape, a shape it called the Dreadnought after a class of British battleships [3].

Fig. 9.6 A Taylor Model 114 Acoustic Guitar (Image courtesy Taylor Guitars, http://www.taylorguitars. com)

In 1931, the Ditson Company was sold and Martin began to make the instruments for themselves. In 1933, cowboy singer Gene Autry ordered a special guitar that went into production as the D-45. It had a 12 fret neck and a body shape unlike the other Martin guitars. While the D-45 was popular, it wasn't representative of the rest of the Martin product line. Indeed, Martin appears to have been ambivalent about marketing it.

The situation changed in 1935 when they introduced the D-28 and its less-expensive companion, the D-18 (analogous to the OM-28 and OM-18). These had 14 fret necks and other features that clearly identified them as Martin guitars, including X-bracing. They also had a more angular shape than the D-45. With their larger body volume, the dreadnoughts had a more pronounced bass response. The D-18, in particular, became very popular and was widely used by country musicians. Elvis Presley started out with a D-18 and traded it for a D-28 with a hand-tooled leather jacket. Figure 9.7 shows a very nice pen and ink drawing of a D-28 by Dick Boak of Martin Guitars.

Perhaps the three most important things that make the D-28 iconic are the body shape, the characteristic deep sound and the X-braced top. Certainly, they are strongly inter-related. Fig. 9.8 shows the interior of a Martin dreadnought guitar. The angular body shape and the X-bracing are clearly visible.

Fig. 9.7 A Pen and Ink
Drawing of a D-28 (Image
courtesy of Dick Boak, http://
www.dickboak.com)

9.4   Gibson L-5

The Gibson L-5 was first produced in 1922 and various versions of it have been in
production ever since. It was the first popular guitar with f-holes rather than a round
or oval soundhole. From a structural standpoint, the original L-5 represented
something of a mid-point between a flat top acoustic guitar and a cello or other
member of the violin family.

The top and back were arched, as are violins, while the sides are were bent into
the familiar shape of a conventional acoustic guitar (like an OM-28). The bridge is
floating – held in place by the force of the strings. Early L-5s had a dark sunburst
finish and a raised black pickguard. Figure 9.9 shows a 1925 model L-5. Archtop
guitars typically have only two soundboard braces, either nearly parallel to one
another or crossing to form a shallow X [6].

The L-5 was adopted by a range of different musicians. One of the earliest
players to become closely associated with the L-5 was Mother Maybelle Carter [7],
a country musician who played with the Carter Family (she was also the mother of
June Carter Cash, wife of Johnny Cash). The L-5 established the form of

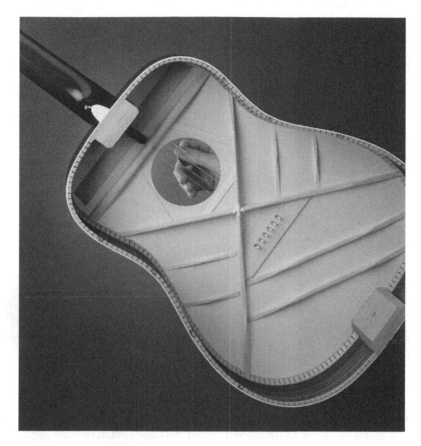

Fig. 9.8 X-Bracing in a Martin Dreadnought Guitar (Image courtesy of C.F. Martin Archives., http://www.martinguitar.com)

generations of archtop guitars. Figure 9.10 shows Bob Benedetto, perhaps the premier modern maker of archtop guitars, with one of his Sinfonietta™ model instruments. This fine modern instrument is clearly a descendent of the L-5.

The L-5 was originally an acoustic instrument, but later versions were fitted with increasingly elaborate pickups, bridges and tremolo tailpieces. Almost all archtop guitars also got a body cutaway. Electric versions of the L-5 and similar instruments became popular among jazz musicians who preferred a body cutaway that allowed better access to the higher frets. Figure 9.11 shows a Gibson L-5 CES with a cutaway and two humbucking pickups.

Note the pickup selector switch on the upper bout near the cutaway and the four knobs (two volume and two tone) below the bridge. The additional mass of the humbucking pickups, metal bridge and potentiometers knobs greatly affects the acoustic properties of the instrument, so it is essentially an electric guitar.

Many manufacturers make archtop electric guitars reminiscent of the L-5, but Gretsch may be the most well-known [8]. Gretsch is well-known for using Bigsby

Fig. 9.9 A 1925 Gibson L-5
(Image courtesy of Elderly
Instruments, http://www.
elderly.com)

vibrato tailpieces. In the jazz world, some of the most prized instruments are archtops made by John D'Angelico (1905-1964) and, later, by his apprentice Jimmy D'Aquisto (1935-1995). While certainly not copies of the L-5, they are quite reminiscent. It's perhaps not surprising that some of the most sought after electric archtop jazz guitars now being made are by Bob Benedetto. Figure 9.12 shows a Benedetto Manhattan with a single neck pickup and the trademark ebony tailpiece.

9.5 Fender Telecaster

So much has been written about the Telecaster that it is hardly necessary to describe its history here [9, 10]. The Telecaster appears here because it was the first commercially successful solid body electric guitar and because it was designed from the start to be easy to make. It has a simple slab body and the bolted-on neck is also milled from a straight blank.

There have been many different Telecaster models since the design was introduced in 1950. It was introduced with a single pickup and originally called the Esquire. A two pickup version called the Broadcaster was introduced shortly

Fig. 9.10 Luthier Bob Benedetto with a Sinfonietta Acoustic Archtop (Image courtesy of Bob Benedetto, http://www.benedettoguitars.com)

afterward. Intellectual property concerns led to the instrument being renamed as the Telecaster. Figure 9.13 shows a typical Telecaster with two pickups and a sunburst finish. Note the large white pickguard covering much of the top bout. On some models, this covers wiring channels on the body, eliminating the need to drill long holes through the body and then run wires.

The Telecaster has been in production for 60 years at this writing and it's tempting to think of the design is essentially static. Indeed, there is certainly nothing wrong with fidelity to vintage designs. However, there is no reason that newer versions of a classic design can't be innovative. A nice example is the Telecaster Acoustasonic as shown in Fig. 9.14.

The Acoustasonic looks like a traditional Telecaster at first glance, but the bridge is clearly different. It is essentially the same size as the traditional metal bridge, but is made of wood. It is also missing the bridge pickup. There also pre-amp controls on the side of the body facing the player. Clearly, this is something quite

different that the traditional Telecaster. This instrument was designed to be a hybrid, capable of producing both representative electric and acoustic sounds while retaining the familiar shape and feel of a Telecaster [11]. The resulting instrument offers the player tonal flexibility not at all possible with the original design.

Some of the unique features of the instrument are visible from the outside and some are not. The body is chambered – it has hollow areas cut into it – that reduce the weight and affect the dynamic response of the instrument. Also, there is a block of spruce under the bridge (and hidden by the ash top plate that covers the body) that changes the mechanical impedance at the ends of the strings. However, some of the most significant changes are in the electronics.

A typical Telecaster has two passive pickups, simple passive tone and volume controls, and a three position selector switch. The Acoustasonic has a single magnetic pickup at the neck position and a piezoelectric pickup under the saddle. They are accompanied by a Fishman Aura® preamp that includes digital processing and software to acoustically image four representative acoustic guitars, called dark folk, bright folk, dreadnought and jumbo. The three position switch allows the player to select the magnetic pickup, the piezoelectric pickup with processing or both. Finally, there is both stereo and mono output available so that the user can send the output of both pickups to the same amplifier or send each to a separate amp.

Fig. 9.12 Benedetto
Manhattan (Image courtesy
Bob Benedetto, http://www.
benedettoguitars.com)

9.6   Fender Stratocaster

As important as the Telecaster was, the next design from Fender, the Stratocaster
was a turning point in the design of electric guitars. The Stratocaster was designed
using both experience from producing the Telecaster and requests from musicians.
The result was an instrument that offered improvements in playability, tonal range
and ease of manufacture.

The slab body of the Telecaster was not always comfortable to play, so the
Stratocaster had deep contours cut into the body. Also the top horn was extended to
move the front strap button to nearly the 12th fret.

The electronics were also significantly changed. The Stratocaster got a third
pickup and an additional tone control (two tone knobs and one volume knob). The
volume knob was placed very close to the bridge pickup so that it could be adjusted
easily with the player's little finger. The instrument was originally supplied with a
three position switch, but it was eventually replaced with a five position switch to
provide two more pickup combinations. Perhaps the biggest change, though, was
the addition of a tremolo bridge.

The bridge was fixed to the guitar only on its front edge. A block extended
through a large pocket cut clear through the body where strong springs held it in

Fig. 9.13 Fender Telecaster
(Wikimedia Commons,
image is in the public domain)

place. Figure 9.15 shows an Eric Johnson signature Stratocaster from both the front and back. The spring pocket cover has been removed to show the springs holding the bridge in place.

While not apparent to the player, some of the most important features of the Stratocaster have to do with ease of manufacture. All the electronics are mounted on the large pickguard so it can be assembled as a separate unit and loaded into the guitar during final assembly. The only electronic component not part of the pickguard is the output jack.

9.7 Fender Precision Bass

Leo Fender's precision bass was not the first solid body electric bass guitar. That distinction probably belongs to Paul Tutmarc, who developed a guitar-like electric bass he called the Model 736 Electronic Bass [12, 13]. His company, Audiovox Manufacturing, began advertising the instrument in 1937. It is unclear whether Leo Fender knew of this instrument or its successor, the Serenader, marketed by Paul's son, Bud Tutmarc. There were also electric upright basses being marketed by several firms in the 1930's. However, none of them had the effect on the

Fig. 9.14 An Acoustasonic Telecaster (Image courtesy of Fender Musical Instrument Corp., http://www.fender.com)

Fig. 9.15 Fender Stratocaster (Wikimedia Commons, image is in the public domain)

Fig. 9.16 Body of a Fender Precision Bass (Wikimedia Commons, image is in the public domain)

market and on subsequent design as did the Fender Precision Bass, introduced commercially in 1951. A modern version is shown in Fig. 9.16. It quickly became popular and has been in production ever since.

The precision bass closely resembles the Stratocaster guitar, which it predates by several years. It had a bolted-on neck milled from a flat blank. It has a single pickup mounted directly to a large plastic pickguard that covers the pickup pocket, the electronics pocket and wiring channels.

Figure 9.17 shows a drawing from patent D187001, filed Jan 6, 1959 and issued Jan 5, 1960. It is titled 'Bass Guitar' and was awarded to Clarence L. (Leo) Fender. The drawing clearly shows covers over the bridge and pickup and a finger rest mounted on the pickguard.

Like other Fender designs, this instrument was and is very widely used and was designed from the start to be easy to manufacture. Viewed from the side, the instrument appears to be almost a slab; the neck is parallel to the body and the body is simply milled from a rectangular blank.

Fig. 9.17 Drawing from
Patent Application D001,
Bass Guitar (US Patent and
Trademark Office, image is in
the public domain)

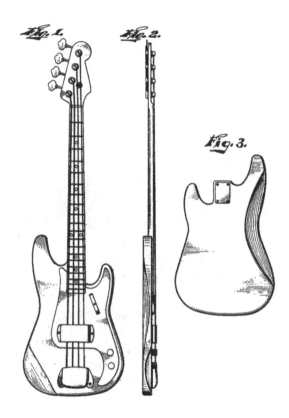

9.8 Gibson Les Paul

The Gibson Les Paul was designed by Ted McCarty, president of Gibson, and Les
Paul, a popular musician and accomplished inventor who was to endorse the new
instrument. It was first sold in 1952 and continued in production until 1960. Due to
popular demand, production resumed in 1968 and has continued since [14].

The Les Paul is about as widely known as the Stratocaster and has greatly
influenced the design of solid body electric guitars. However, it is conceptually
almost the opposite of the Strat. It doesn't have the simplicity of the Stratocaster
and is more difficult to manufacture. The Les Paul is reminiscent of an archtop such
as the Gibson L-5. The top is elegantly arched and the neck is glued to the body
(a 'set neck'). The neck is not parallel to the plane of the body, but set at an angle as
shown in Fig. 9.18.

The pickups are set into routed pockets while the electronic pockets are cut into
the back of the instrument. The selector switch is not mounted with the tone and
volume knobs (one pair for each of the two humbucker pickups), but at the top of
the upper bout. This requires a wiring passage to be cut diagonally through the
entire body. Even the neck is more complicated, having an angled headstock,
reminiscent of acoustic guitars, rather than a flat headstock like the Fender electrics.

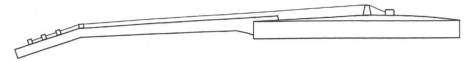

Fig. 9.18   A Guitar Neck Set at an Angle to the Body

Fig. 9.19   Gibson Les Paul (Wikimedia Commons, image is in the pubic domain)

The result of these complications is an elegant, attractive instrument that has been popular among several generations of musicians. A modern Les Paul is shown in Fig. 9.19.

References

1. Osborne N, ed. (2002) The Classical Guitar Book: A Complete History, Backbeat Books
2. http://mysite.ncnetwork.net/resr7g3w//, last visited June 3, 2011.
3. Johnson R and Boak D (2008) Martin Guitars: A History, Hal Leonard Books
4. Mott RL (2007) Applied Strength of Materials, Prentice Hall
5. Taylor B (2011) Guitar Lessons, Wiley
6. Benedetto B (1996) Making an Archtop Guitar, Centerstream Publications.
7. Zwonitzer M and Hirshberg C (2004) Will You Miss Me When I'm Gone? The Carter Family and Their Legacy in American Music, Simon and Schuster.
8. Bacon T and Day P (1996) The Gretsch Book – A Complete History of Gretsch
9. Bacon T (2001) The History of the American Guitar, Outline Press
10. Kelly M Kelly P and Foster T (2010) Fender: The Golden Age 1946-1970, Cassel.
11. Wagner J (2010) Fender Acoustasonic Telecaster Electric Guitar Review. Premier Guitar, July
12. Blecha P (1999) Audiovox #736: The World's First Electric Bass Guitar. Vintage Guitar, March.
13. Tutmarc.tripod.com/paultutmarc.html, retrieved March 18, 2011
14. Duchossoir AR (1981) Gibson Electrics, Hal Leonard Publishing